NETWORKS IN SOCIETY

LINKS AND LANGUAGE

NETWORKS IN SOCIETY

LINKS AND LANGUAGE

edited by
Robert Stocker
Terry Bossomaier

PAN STANFORD PUBLISHING

Published by

Pan Stanford Publishing Pte. Ltd.
Penthouse Level, Suntec Tower 3
8 Temasek Boulevard
Singapore 038988

Email: editorial@panstanford.com
Web: www.panstanford.com

British Library Cataloguing-in-Publication Data
A catalogue record for this book is available from the British Library.

Networks in Society: Links and Language

ISBN 978-981-4316-28-6 (Hardcover)
ISBN 978-981-4364-82-9 (eBook)

Printed in the USA

Vale John Carroll

It is with deep regret that we acknowledge the passing of our friend and colleague Professor John Carroll in October 2011, and in recognition of his valuable contribution to the research community, we dedicate this publication to his memory.

Contents

Preface

The growth in our understanding of networks, especially human social networks, has been quite remarkable over the last decade. But equally remarkable is how the networks themselves have been evolving, driven in part by the many new tools in cyberspace. Along with these new communication tools come increasing numbers of changes to language and lexicography. English spelling in text messages is nothing like the spelling we learn in school, so far at least, while groups differentiate themselves more and more by linguistic twists and turns, new words, slight variations in grammatical usage—the *so not cool* phenomenon.

Although this book can only provide a snapshot, since this is the forefront of social evolution in the 21st century, we present a broad range of ideas from several fields of research endeavour (and we thank the authors for their respective contributions).

We discuss social networks and their integration with communication and language. Although accessible to a wide audience, it contains sufficient technical detail to serve as a starting point for advanced undergraduates and postgraduates and reflects the content of the 5th Biennial Complex Systems Research Summer School held at the Centre for Research in Complex Systems (CRiCS), Charles Sturt University, Bathurst, Australia, in December 2009. The seven chapters of the book cover three broad areas: technical fundamentals, complexity and social networks, and communication and language. Chapter 1 explores the development of some contemporary models, exposing key ideas in the relationship between social networks, patterns of connection and language.

Since the 18th-century mathematician Leonhard Euler formulated and solved one of the first classic problems in graph theory, practical applications of this esoteric mathematical field have steadily emerged, but social networks have caused rapid expansion. Some understanding of the basic mathematics is an essential starting point. Concepts and results belong here, but proofs and finer points need to be sought from some of the excellent textbooks now publicly available. Alongside the mechanics of graph and network theory, two other topics are covered. Firstly, we need to know how information about networks is collected, from surveys, interviews, email and other techniques, some quite new. Secondly, visualization is enormously important in understanding network structure and the book introduces some of the well-established software for mapping abstract network representations to something we can look at and grasp intuitively. Chapter 2 identifies important metrics associated with graph theory and network theory and discusses useful tools that assist in visualizing network structures. Chapter 3 provides the links between social structure, cognitive performance and language in social groups.

The field of social networks is already too large to cover in a book of this size, and some framework is needed to make a coherent selection. The framework we use derives from complexity and evolution. The key to interest in social networks is their distinctive structures, such as Milgram's six degrees of separation and Watts' small world ideas, now, as they say, "a major television series"! Chapter 4 explores key issues in complex social structure and identifies important concepts that will occupy future research programs in social complexity. But these structures are dynamic, and there are many new ideas to consider in how social networks evolve over time.

Evolution brings us to the third major theme, communication and language. Language has always defined social groupings, from entire language across frontiers to changes in dialect within an ostensibly homogeneous country. Physical co-location is no longer a prerequisite for social structures. Email, text messaging, Twitter electronic communication tools give us near-instantaneous contact across the globe, while FaceBook, LinkedIn and others operate across geographical boundaries. Many readers are likely to "Tweet" their way through the day, new usages that reflect the tight coupling of language, new media and social structures. Do these new technologies reflect ongoing changes in the size, strength of interaction and geographical distribution of our personal networks? All these issues make the evolution of social networks one of the most fascinating domains of current research. Chapters 5 to 7 discuss practical applications of the ideas explored in earlier chapters with specific emphasis on business modelling, natural resource management in sensitive locations, and the evolution of social interaction resulting from new information and communication technologies.

Our intent is to raise your interest in social networks in general, the importance of language in the emergence and maintenance of networks in particular, the integral patterns of interconnectivity, and the means to measure and model social structure. The extent to which the contributing authors' knowledge and experience forms the basis for your own investigation of the phenomenal world of social structure will be our measure of success for this work.

In the end, it is our hope that you will be stimulated to build on these writings to explore more deeply the ideas, concepts, perceptions and thoughts that personally appeal to you. That these works become the staging point for further investigation would be our best return on investment. May your connections prosper!

Robert Stocker
Terry Bossomaier
Winter 2013

Chapter 1

Social Networks: Towards General Models

Robert Stocker
School of Engineering and Information Technology,
University of New South Wales at The Australian Defence Force Academy,
Northcott Drive, Canberra ACT 2600, Australia
r.stocker@adfa.edu.au

Aggregation of entities with common characteristics appears to be a universal phenomenon associated with complex systems. Survival, reproductive or safety behaviours are evident in most plant and animal species in which integral "social" behaviours have also emerged. Trees, and vegetation, of similar species tend to "clump" or "cluster" together, birds flock, monkeys groom, cattle and sheep herd, fish school, and so on. In humans, resulting from or concurrently with the evolution of larger brains and an innate capacity for language, social behaviour is observed in the formation of networks of association that transcend these somewhat instinctive elements of our lives.

1.1 Introduction

What is it that enables human social behaviours?

Networks in Society: Links and Language
Edited by Robert Stocker and Terry Bossomaier

ISBN 978-981-4316-28-6 (Hardcover), 978-981-4364-82-9 (eBook)
www.panstanford.com

Humans are systemic. Internally we comprise organic, biological, neural, digestive, respiratory, skeletal, muscular, vascular and other systems. Externally, we interact in social systems (as part of our world environment) that depend on spatial and social proximity, behaviour, capability, knowledge of concepts and ideas, beliefs and values, and personal identity.

Humans are modellers—we are wired to manipulate symbols. We are accomplished builders of symbolic representations of what we perceive through interaction with our personal systems, other human beings, and the environment in which we exist. The mental maps (models) of our experience that we construct in our brains are symbolic representations or abstractions of what is "real" and, in fact, impoverished models of the world that surrounds us [17]: we are "victims of an illusion" [42 p. 4].

Humans are communicators—we are wired for language. It gives us the capacity to communicate our perceptions of our worlds to others as oral and written symbols of what we see, hear, feel, taste and smell—our sensory mechanisms.

Humans prefer order to chaos, but order with a degree of variation to it. Chaos to us represents noise, that is, any pattern with which we interact that we do not understand. As "voracious consumers of patterns", we are "best at filling in the blanks and making assumptions" [38].

How is this so?

1.2 Social Networks

No person is an island. Individuals form into networks that are established by linking together through various means. Family, work, recreation, clubs, associations, electronic communication and other social "clusterings" occur between people with some common interest. However, people may also be connected for other reasons, such as health conditions, common behaviours and language, or other attributes. They are often linked to key prominent individuals who are recognized as "leaders" (Fig. 1.1).

How big are these networks? From an evolutionary perspective, we emerged from a primate ancestry where grooming was the

Social Networks

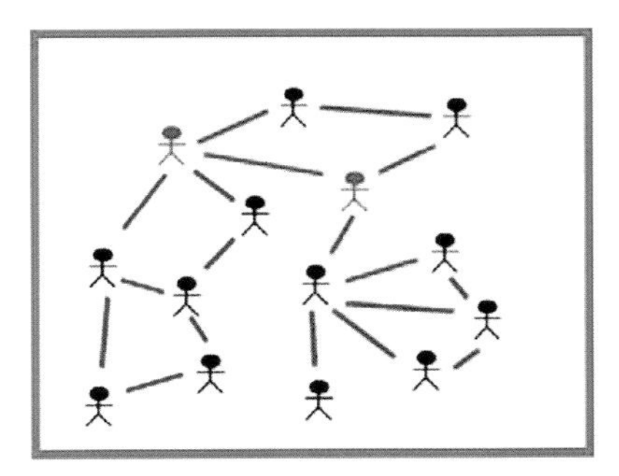

Figure 1.1 Graph of a social structure that shows subgroups (red and blue links) linked by key agents (green nodes).

principal means of communication and social group size comprised around 30 to 40 members. Coincidentally, as our need to accommodate larger group sizes increased, our capacity for language developed, our brain size increased and group sizes grew to around 100 to 160 [20–22]. In general, a human will have intimate interaction with three to six persons—that is, these people will be in quite close contact several times each week. At the next level of connectivity, the average human will have some 10 to 20 close friends with whom contact is strong but less frequent. At the next level, the average human will have up to 150 acquaintances that form more casual associations. Finally, around 1000 "known" other people form the least strongly connected component of an individual's personal social network [20–23, 50, 51].

Interestingly, there appears to be a "tipping point" or phase change [19, 30, 35, 39] in conversational social group size as the number of members approaches six. At this point the group will typically break into two or more separate groups, suggesting that, for conversational (linguistic) efficiency and effectiveness at least, 5 is the optimal number of members.

Social structure comprises networks of connected individuals who for various reasons will form links of various strengths between each other. Interaction, regardless of network size, types of individuals, or numbers of links to other individuals, can be described

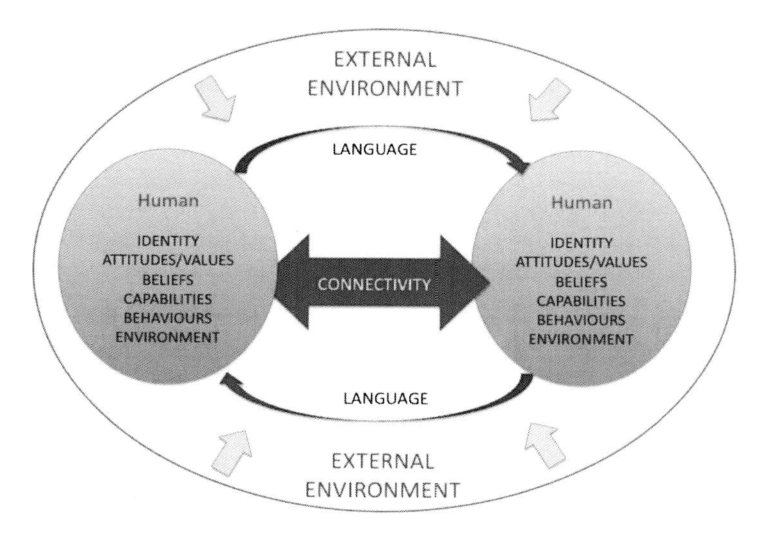

Figure 1.2 An overview model describing the process of social interaction between individual human participants, emphasizing key human characteristics and the importance of language in connectivity patterns between interacting actors in a dyad.

as multiple numbers of single links of varying intensity between numbers of pairs of individuals (a dyad). The connectivity and interaction involves the characteristics of each individual (see Section 1.7), communication between them, the strengths and direction of the ties between them, the levels of influence each has over the other, and the cultural and social constraints that are included in the environmental feedback in which interaction takes place (Fig. 1.2).

So we can consider linked groups of linked individuals and thus increasingly larger social networks. The same patterns of connectivity and interaction occur, but here we can observe patterns of behaviour and characteristics of sub-groups or the whole network structure. Thus, networks themselves can be categorized by their specific structural characteristics, for example as random [24], small world [8, 48, 49], epidemiological [2–3], scale free [7], hierarchical [32, 45–46] and so forth. Such networks are characterized by various metrics, for example, clustering coefficient, centrality, assortativity and the like (see Chapter 2).

How is it then possible to rigorously investigate social structure?

1.3 Social Networks Research

Social networks research is based on principles and characteristics that are relevant to the area. These principles have emerged and combined from multidisciplinary sources over time and encompass a range of methodologies and technologies that give social network research the ability to identify and examine patterns of human social behaviour (e.g., interaction and connectedness) referred to as social network analysis (SNA) [37, 44]. In turn this allows for the development of models that are explanatory, predictive and above all testable by other researchers.

What constitutes social network research—that is, its system of knowledge, its methodologies and its technologies that lead to credible general models?

1.3.1 *How Do We Know What We Know: The System of Knowledge?*

The epistemology of social network research [29], that is, how we know (analysis) what we know, is reflected in its relationship to the fields of research from which it has emerged. Social network research is principally structuralist. It creates links to existing knowledge in order to expand knowledge in the field. It has evolved from well-established research disciplines such as complexity, biological, sociological, psychological, cognitive, mathematical, physical and other sciences that support the principles and practice of social network research [31–32, 34].

1.3.2 *How Do We Apply What We Know: The Processes and Procedures for Applying That Knowledge?*

Research in social networks applies three systems of transformations that help us to develop models and to explain their purpose (adapted from [18]):

1. *Inductive transformations* allow us to examine patterns in, and build models of, the social network world around us. Here we look for high-level patterns such as concepts, ideas and universal

principles about social connectivity and behaviour that can be identified as the "deep structure" of social network research.

2. *Deductive transformations* allow us to explain, and act upon, our perceptions and models of the social network world around us. Here we expand the "deep structure" general concepts and ideas into specific "surface structures", such as specific words, descriptions, language and behaviours that facilitate communication between those involved in social network research.
3. *Abductive transformations* allow us to transition between one deep structure and another (homomorphism: creative and imaginative constructs)—that is, to explore novel ideas and principles about social network research—and one surface structure and another (isomorphism: pattern similarity constructs), that is, common linguistic descriptions relevant to social network research.

1.3.3 *How Do We Interpret What We Know: The Tools and Technologies That Inform Our Application of the Knowledge?*

The scope of social network research that encompasses high-level concepts, ideas and universal principles, the language and behaviours of social network researchers and mediation between them allows for the exploration of theoretical and practical aspects of social networks. It encourages the development of new theories and their communication between researchers with a common interest in social constructs. It provides within the common language a range of metrics to describe social structures (see Chapter 2). It provides tools for visualizing and interpreting data gathered from real world and simulated social investigation (see Chapters 4, 5, 6 and 7). It provides techniques for developing models of social structures. It allows for experimentation in situations where real-world exploration may be difficult or impossible [27]. It provides artificial societies based on computational simulation and serious games research [28].

1.3.4 *Modelling*

Fundamentally, researchers are concerned with the design, construction, implementation and evaluation of models that serve to inform, explain, and predict characteristics and behaviours of particular complex natural phenomena—in this case the networks that result from social interaction between human beings.

There are, of course, many questions about the emergence and maintenance of social network structures that remain unanswered [25–26]. These questions arise from our limited but evolving knowledge of the complexity associated with (to name but a few): the human organism itself, the processes of interaction between humans, the patterns of connectivity between individuals and groups, the environment(s) in which humans interact, the mechanisms of feedback from members of a social structure together with the environment in which those structures exist, the structure and function of the human brain, the process of cognition, the communication between individuals and networks, and the nature (i.e., the syntax and semantics) of formal language(s) used in communication.

Formulating a rigorous scientific modelling process for researching social systems thus becomes a process of selection from a confusing array of variables from which to choose a pathway for the examination of specific social phenomena. Reductionism must, of necessity, serve an important function in the examination of such complexity. However, the rise of significant computational power and access mitigates this necessity somewhat. Thus, more deeply complex models can be developed to examine the influence of a wider variety of factors.

From the earliest forays in social structure research, scientists have used models to represent their ideas about social structure and behaviour (see, e.g., [44, 47]). One model in particular has emerged as a consistent method for representing social structure, the directed graph in which nodes represent individuals, lines between nodes represent connection, arrows on the ends of lines to represent the direction of flow of communication and/or resources, and numerical values on lines to represent strengths of connections between nodes (see Chapters 2 and 3).

Recent work on models of the emergence of stable network structures (see, e.g., [11, 25–26]) demonstrates the relationship between context and the consequence of conscious actions by actors.

Serious games and simulations (see Chapters 5, 6 and 7) applied to research are a recent development that holds promise in unlocking the hard problems in the study of social network complexity [28].

However, what contributes to the ability to form social networks? Are we able to identify elements that provide evidence as to our systemic capacity for modelling, communicating and pattern consuming?

1.4 Sensory Experience

Immediate contact with internal and external experience is through our sensory receptor systems. Our sight, hearing, feeling, smell, and taste reception are sensitive to information that is specific for each sensory receptor system. Input consists of visual (V), auditory (A), kinaesthetic (K), olfactory (O) and gustatory (G) primary modality distinctions [10, 17–18]. However, these sensory systems have limitations: for example, through our visual receptor system, we can receive neither the infrared nor the ultraviolet ranges in the spectrum of electromagnetic waves known as light—yet they exist.

Each primary modality distinction has intrinsic sub-modality attributes. For example, our visual receptor system (primary modality) is uniquely structured to accept data that discriminates between, colour, size, shape, movement, distance and similar distinctions (sub-modalities). Likewise, our auditory receptor system is sensitive to pitch, volume, timbre, tempo, etc. [6, 17–18]. Human sensory mechanisms primarily produce information about changes in our environment rather than registering constant aspects of experience: a difference that makes a difference.

However, we tend to pay most attention to visual, auditory and kinaesthetic {V, A, K} cues although our genetic make-up and experience influence such selective attention. It is reasonable to assume that in specific circumstances individuals will consistently use particular sensory mechanisms to gather data about the environment in which they are operating. For example, a wine taster would tend to pay more attention to olfactory and gustatory input signals while evaluating a particular wine. Visual cues appear to be used most by humans (~60%), with auditory (~20%) and kinaesthetic (20%) used less frequently [10].

Selective attention to our internal or external experience is, then, governed by our personal preferences for specific sensory components, the intensities of the modalities and sub-modalities of an experience—that is, our reception is guided by our perception. For example, because of previous experience, we may prefer to pay attention to the brightness and speed of movement, together with the clarity and loudness of sounds associated with a specific current experience.

1.5 Towards a Model

A preliminary mathematical model for sensory processing and the production of our mental models of our experience of reality can be proposed.

Our perception (p) of an event or experience (ε) can be expressed as a set (Eq. 1.1) of the levels of intensity (l) of our internal (i) and external (e) sensory input data (SD):

$$p(\varepsilon) = \{\mathrm{SD}[1]_l, \ldots, \mathrm{SD}[5]_l\}^i{}_e \qquad (1.1)$$

Bandler and Grinder [4, 5] identify this selectivity as "neurological constraints", unique to each person, translating raw sensory input data into bioelectrical impulses that are transmitted via sensory channels. Our bioelectrical maps are also modified by filters called "social constraints", referring to a code of symbols that represent our cultural and linguistic influences on the sensory representation of an experience (see Section 1.6). These neurological (n) and cultural/sociological (s) constraints act to process this perception to form a number ($1 - m$) of deep structures (DS) representations that are an ecological equilibrium function (E) with a number ($1 - k$) of related existing internal mental maps (Eq. 1.2).

$$\mathrm{DS_1}^m = E\{p(\varepsilon)^n \Leftrightarrow p(\varepsilon)^s \Leftrightarrow \Sigma_1{}^k[\{\mathrm{SD}[1]_l, \ldots, \mathrm{SD}[5]_l\}^i]\} \qquad (1.2)$$

This assimilation of bioelectrical representations is accomplished by the self-organizing capacity of the neural networks of the brain and the top-down and bottom-up processing that occurs throughout the cortex. Consciousness, awareness and selective attention are necessary to match and mediate the neural net activity that produces equilibrium.

1.6 Cognitive Structure

Are we hereditarily "wired" for specific sensory functional preference? Such focus of attention would result in the emergence of a specific network of connected neurons against the background of all other possible neural networks. The "noise level" or level of difference between an emergent neural net and the background is the key to the focus of attention needed [1].

At synaptic junctions bioelectrical impulses are converted to biochemical signals and transmitted across synaptic gaps to be "reconstituted" as new bioelectrical impulses. They are stored in various areas of the brain and connected via cortical association areas [1]. Neural connections, representing bioelectrical maps, depend on activity where co-activated neurons develop into cell assemblies. Connections are strengthened or weakened by the amount of activity associated with them [33]. The neural connections that form or decouple are dependent upon our perceptions of personal experience and the cultural or social filters of our environment. Thus, for a single experience, different individuals may trigger quite different bioelectrical impulses producing significantly different perceptions of that same event.

Cognitive function is dependent upon individual neural organization for processing internally or externally generated sensory experience, and consists of varying intensities of visual, auditory, kinaesthetic, olfactory and gustatory modality and sub-modality inputs. Cognitive representations of sensory inputs that are stored as "memory maps" rely on cellular mechanisms, neural structures and modifications to strengths of neural connections [1, 33, 36]. For example, strong neurophysiological evidence suggests the existence of auditory memory processing modules for sub-modality attributes of frequency, intensity and duration.

The resulting memory maps are known as "deep structure" [13–14, 42]. "Surface structure" is the translation of deep structure symbolic representation in to linguistic representations [10]—see Section 1.7.

1.7 Cognitive Processing and Logical Levels

The same physiological, neurological, sociological and cultural constraints that allow us to construct symbolic representations of our experience with reality also give us another unique capacity.

It would be difficult to imagine that a human being could attend to every one of the billions of bits of sensory data that impacts the sensory receptors every second of that being's existence. We simply cannot accommodate the sheer volume that accompanies interaction with others and the environment in which experience occurs. So we have evolved to be selective in our attention to those cues that enable our survival and continuation of the species. We are able to "store" bioelectric and cultural adaptations or reconstructions of reality through the same actions that provide us with an impoverished model of reality.

How is that accomplished?

Generalization, deletion and distortion are three processes by which we transform sensory input through surface structures into deep structures and those deep structure concepts and ideas into communicable or linguistic representations as surface structures [5, 13–14, 17, 40–41]. Different surface structures can be representations of common deep structures (Fig. 1.3).

Thus the human brain is good at cutting out the irrelevant, noticing a lot more than we think it does, and actively hiding the real world from us [38]. This process is generally referred to as "chunking".

Thus, chunking patterns become automated and we primarily operate on these as we interact in our real worlds. When our chunks let us down, that is, something behaves differently from our expectations (learned behaviour), conscious thought (which is really

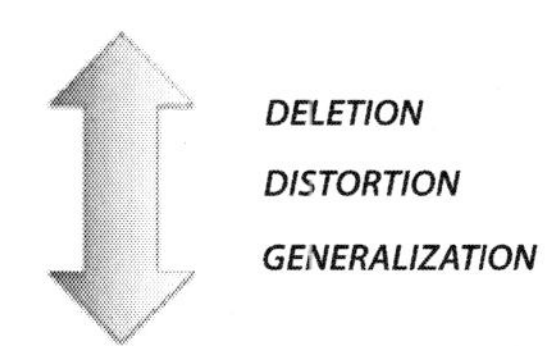

Figure 1.3 Generalization, deletion and distortion are the filters that transform deep structure to surface structure, and vice versa (adapted from [10, 18]).

inefficient) takes over and our reactions become slowed by orders of magnitude [38].

How can generalization, deletion and distortion be represented?

Let us assume a population (P) of agents (A) initialized with up to (N) sensory tuples (T) each that represent internal memory of experiences. Thus, a social group (G) can be expressed as

$$G = \{A[a = \{T[n = \{\mathrm{SD}[i]_{i=1}{}^{5}\}]_{n=1}{}^{N}\}]_{a=1}{}^{P}\} \qquad (1.3)$$

From the preliminary sensory processing model described in Section 1.6, we have, for each external event, sensory data in the form of a tuple, which can be represented as $\{\mathrm{SD}[i]_{i=1}{}^{5}\}^{e}$ and the intensity of each sensory signal $\mathrm{SD}[i]$ can be given a value as a probability of its level of influence (e.g., 0.15 or 15%). Generalization, deletion and distortion are internal processes for each individual (or computational node in a simulation) and we can represent them as types of functions, where the internalized (I) representation is

- Generalization (an averaging function)
  Trending of an agent's incoming (external) tuple values towards the average of all internal tuple values:

  FOR any agent A :
  $$\mathrm{Average}\{\mathrm{SD}[i]_{i=1}{}^{5}\} = \Sigma_{n=1}{}^{N}\{T[n\{\mathrm{SD}[i]_{i=1}{}^{5}\}]\}/N \qquad (1.4)$$

the difference between the external tuple value and the average tuple value is:

$$
\begin{aligned}
&\Delta\mathrm{SD}[i]_{i=1}{}^{5} = (\{\mathrm{SD}[i]_{i=1}{}^{5}\}^{e} - \mathrm{Average}\{\mathrm{SD}[i]\}_{i=1}{}^{5})\\
&\mathrm{FOR}\ i = 1\ \mathrm{TO}\ 5\\
&\mathrm{IF}\{\mathrm{SD}[i]\}^{e} < 1.00\ \mathrm{AND}\ \Delta\mathrm{SD}[i]\\
&\quad < 0, \{\mathrm{SD}[i]\}^{I} = \{\mathrm{SD}[i]^{e}(1 - \alpha\Delta\mathrm{SD}[i]\})\\
&\quad > 0, \{\mathrm{SD}[i]\}^{I} = \{\mathrm{SD}[i]^{e}(1 + \alpha\Delta\mathrm{SD}[i]\})\\
&\quad = 0, \{\mathrm{SD}[i]\}^{I} = \{\mathrm{SD}[i]\}^{e} \qquad (1.5)
\end{aligned}
$$

where α (a randomly seeded threshold constant) < 0.10.

- Deletion (a minimizing function)
  - Random reduction to zero of an agent's external tuple member's value(s) such that:

$$
\begin{aligned}
i &= \mathrm{RAND}\{1, 5\}\\
\{\mathrm{SD}[i]\}^{e} &= 0 \qquad (1.6)
\end{aligned}
$$

  - Reduction to zero of any selected agent's tuple member(s) value(s) by sensory selectivity such that (from [Eq. 1.4]):

$$
\begin{aligned}
&\mathrm{FOR}\ i = 1\ \mathrm{TO}\ 5\\
&\mathrm{IF}\ \Delta\mathrm{SD}[i] < 0, \{\mathrm{SD}[i]\}^{e} = 0 \qquad (1.7)
\end{aligned}
$$

- Distortion (an evolutionary function)
  Random mutation (swapping) of tuple member's value(s) such that

$$
\begin{aligned}
&\mathrm{FOR}\ k = \mathrm{RAND}\{1, \mathrm{P}\}, [l = \mathrm{RAND}\{1, 5\}; m = \mathrm{RAND}\{1, 5\}]\\
&\{\mathrm{SD}[l]\}^{k} = \{\mathrm{SD}[m]\}^{k}\ \mathrm{AND}\ \{\mathrm{SD}[l]\}^{k} = \{\mathrm{SD}[m]\}^{k} \qquad (1.8)
\end{aligned}
$$

The various constraints that impact on our ability to perceive the reality of our environment are integral to the process of transformation of perceptions from surface to deep structures in the human brain.

Dilts [18] suggests that progressing from surface structure to deep structure involves progressing through a series of "logical levels" (Fig. 1.4). These logical levels are generic in that they can be applied to individuals, groups, networks, organizations, cultures and nations. Each level (Local Environment, Behaviours, Capabilities,

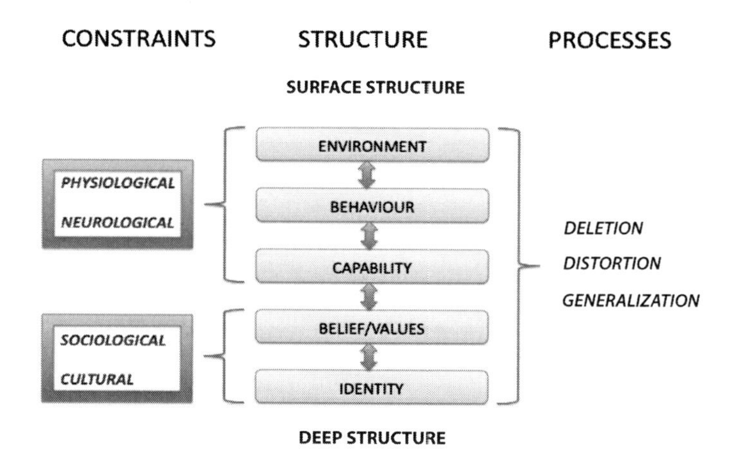

Figure 1.4 Physiological and neurological constraints relate to the processing of perceptions through logical levels from surface to deep structures (adapted from [18]).

Beliefs, Values, Identity, Vision, Spirit, Global Environment) is successively deeper than the one before it.

Koster's [38] three levels of processing also support the transition of subjective experience to deep level "programming":

- Conscious thought is mathematical and responsible for assigning values, making lists, etc.
- Packaging or chunking or "common sense" is integrative, associative and intuitive.
- Automating (the autonomic nervous system) encompasses reflex and trained response, although some conscious control can be applied, for example, modifying breathing and heart rate. Trained responses enable the building of a "library" of chunks through repeated practice.

Interestingly, dreaming can be seen as the equivalent of "burning" neural pathways by the intuitive part of the brain to turn patterns into something that fits in with the existing context of the chunks that we already know and use (see Section 1.7).

Our perceptions and interactions with the environment are part of our experience that is closest to "the surface". Coordinating and managing behavior requires the mobilization of deeper structures

> in our neurology. Our capabilities organize and coordinate our behaviors, involving less concrete but deeper processes. Beliefs and values are the underpinnings of our capabilities and behaviors. They are more difficult to express clearly and specifically at the surface, but they influence us at the deepest level. Identity is a very deep set of relationships that can best be described as a "function" analogous to the equations used to create a mathematical fractal [18].

Thus the deep structure of an individual's identity is evidenced in surface behaviours through transformations at each level that are influenced by the processes of generalization, deletion and distortion [10].

What are those surface behaviours?

There are two principal elements. Verbal and non-verbal behavioural "cues" are indicators of the types of processes taking place as transformation between the logical levels progresses from a deeper level to a level closer to the surface. These two types of cues (verbal and non-verbal) are parts of the total communication "package" evident as people interact in various social structures (dyads, triads, groups, etc.). Verbal behaviour refers to the linguistic distinctions that people use to communicate their experience. Non-verbal behaviour refers to the physiological distinctions that often, but not necessarily always, accompany verbal interaction.

Linguistic cues may include distinctions that define specific sensory processing. For example, "That rings a bell" suggests some sort of auditory processing, or "I see your idea" suggests visual processing and "I can't quite get hold of that" indicates kinaesthetic processing. At this high level these distinctions are called primary modalities and refer to the five principal senses by which we interact with our environment. Refinements of these distinctions are called sub-modalities [6] and offer further clarification of the primary modality processing that is occurring. For example, primary modality visual descriptions may be enhanced by sub-modality descriptions that include referents to size, colour, shape, movement and the like.

Physiological cues include movements, positioning and directing attention towards parts of the body. Well-known instances, such as folding the arms/crossing the legs (body language), that demonstrate non-acceptance are gross indicators of mental processing. Others such as pointing to body parts or touching parts of the head may indicate parts of the brain being used to process thinking. More subtle indicators, such as eye movements, level and depth of breathing, facial expressions, face and lip colour and size changes, etc., can in combination with linguistic cues provide pointers to levels of processing from deep to surface structures [10, 17].

Thus the transition from sensory input to deep structure symbolic representation undergoes processes that incorporate biological, chemical, electrical, personal and social "conditioning" to give humans the capacity to interact with their environment and their perceptions of the "real world" and to effectively and efficiently communicate those representations to other humans through language.

1.8 Language

Human beings are natural communicators. We are wired for language. We cannot not communicate [10, 38, 41 pp. 427–433].

Language, as a three-part system (word formation, concept representation and a process of mediation between the two), compresses concepts of events or experiences that have gone through primary and associative processing and communicates them subsequently in an efficient manner. It is implicit in this view that language has an operational function on cognitive processes in the brain [16].

Attempts to deal with the idea of language must first discuss the concept of "mind". The symbolic nature of language comes from the way the mind processes what it encounters from both external sensory input (provides environmental data) and data that is internally generated. The mind processes this data and then is able to conjure up other sets of symbols that are used to transfer information to other minds that can interpret and understand those sets of symbols, that is, language. Such discussion must also consider the nexus between brain and mind, that is, the relationship between the physical entity and the higher order functions that it performs.

The brain is fundamentally an information processing system. The physical entities (neurons, axons, dendrites, synapses, etc.) all combine to reliably, accurately and efficiently propagate information. Physical changes in the format of the information signal (e.g., electrical to chemical) do not change the information itself. Thus, the representation of the information, that is, its symbolic state, may change, but what that medium carries does not. Such representation comes in four forms [42]: visual image, phonological "tape loops", grammatical structures arranged in hierarchical trees, and "mentalese" or conceptual (symbolic) knowledge.

Pinker [42 p. 21] provides a useful definition of the mind:

> The mind is a system of organs of computation, designed by natural selection to solve the kinds of problems our ancestors faced in their foraging way of life, in particular, understanding and outmaneuvering objects, animals, plants and other people.

Thus, the mind is what the brain, in part, does: it makes us see, hear, feel, taste, smell, think, choose and act; it computes; it processes information; it is modular; its modules engage in specific processing that relate to real-world interaction; and its module logic is genetically programmed.

The computational theory of the mind [15] suggests that our identity, beliefs, attitudes, values, capabilities and behaviour can be thought of as information that is encapsulated as configurations of symbols in the patterns of connectivity of neural physiology in the brain. Neurons in the brain are essentially symbolic logic-processing units that are linked in parallel by axons and dendrites that form intricate patterns of networks, or networks of connected networks. It may be that connectivity is hierarchically structured (as are most complex systems) in symbolic layers corresponding to logical levels (see Section 1.7).

The mind comprises specialized components in order to be able to solve specialized problems. These mental organs, just like our physical organs, have evolved into specific structures that are designed to accomplish specific tasks. Systems of mental organs are not distinct plug-and-play modules, but rather general areas of connectivity, spread across areas of the neocortex, to perform functionally specific processing. Certainly, modern technology and research has identified areas of the brain where specific types of processing take place [1, 15–16, 36, 41, 43]. It is well known that, at least in the sensory areas of the brain, genetic coding (or "recipes") sets the pattern for cell aggregation, molecular migration and elementary connection to appropriate other cells during embryonic development. However, it is specific neuronal activation that brings precision to these patterns of links and establishes the fine-tuned wiring patterns that produce functional modules in the brain.

Concurrent activity in the language regions of the brain (and the sensory and association cortices) leads to assemblies that include neurons from various cortical sites. Connections are strengthened or weakened by the amount of activity associated with them. The neural connections that form or decouple are dependent upon our perceptions of personal experience and the cultural and social filters of our environment.

The symbolic representations that we use to communicate our perceptions of experience and know as language [40] focus our attention towards external as well as internal environments. That language has influences on specific brain areas has been demonstrated by the emotional behaviour instigated via the hypothalamus and amygdala following verbal cues. It is of interest to note that olfactory or visual cues can cause the same emotional response, suggesting cortical interconnection between the association cortices involved with different modality processing.

Steven Pinker [41–43] suggests that language is instinctive, that we are wired for language, that language is integral to the workings of the mind and that language is a window to human nature—in fact, it is the operating system of the brain [12, 18, 43], a further reflection of the computational theory of the mind. The language of the mind is "mentalese", a sort of internal set of symbols that carry semantic, syntactic, contextual and content data across the networks of the brain.

Mentalese comprises ordered sequences or patterns of sensory representations that emerge as "strategies" for communicating our experiences to others. The "wrapper functions" of generalization, deletion and distortion apply bidirectionally across logical levels, thus suggesting a hierarchy of strategies that are relevant to each level. Such strategies emerge from the strengths of sensory components of representations of the experiences that we "store" in our minds, and they can be described by a "systemic" flow. For example, a strategy that takes an external experience (visual external V^e), constructs an internal visual representation ($V^{i,c}$), that can then be internally "sound-checked" for accuracy (auditory digital internal $A^{i,d}$), until a sense of feeling right (K^+) before communicating ($A^{e,c}$) that experience to others might be represented as [17]:

$$\begin{array}{ccccccccc} V^e & \Rightarrow & V^{i,c} & \Rightarrow & A^{i,d} & \Rightarrow & K^+ & \Rightarrow & A^{e,c} \\ & & \Uparrow & & & & K- & & \\ & & & & & & \Downarrow & & \\ & & & \Leftarrow\Leftarrow\Leftarrow\Leftarrow & & & & & \end{array}$$

where the loop structure shows iteration until a positive feeling is generated.

Language is therefore closely linked with an individual's perceptual, conceptual and neurophysiological structures, and therefore any identifiable characteristics of a specific language used by a particular individual will be indicative of processing in those structures, suggesting a dynamic and recursive nature.

The physiological production of language, that is, the physical structures that allow the utterance of combinations of phonemes that humans can pronounce, plays an important role in the process of communication between them, as does the cultural and social environment in which those phonemic combinations are produced. The mouth, tongue, vocal chords, trachea, lungs, etc., are the mechanisms that enable the transfer of personal social and cultural "mentalese" symbols (each person's understanding and interpretation of their "reality") between individuals who possess similar phonetic combinatorial capacities, thus accommodating the differences in language across different societies and cultures.

Thus we progress to a final general model. As beings with an innate capacity for language and communication, we are adept at the physical process(es) of combining phonemes into syntactic and semantic appropriate linguistic frames to represent our interpretation/perception of "reality". This representation begins as sensory input in the form of visual, auditory, kinaesthetic, olfactory

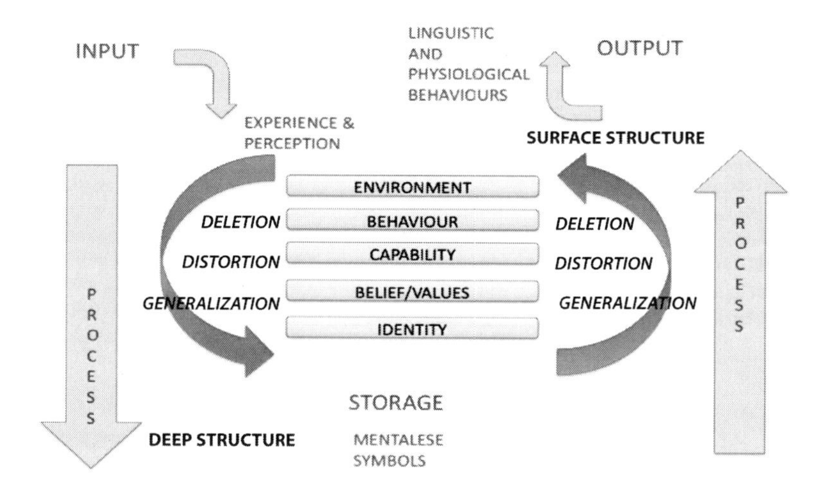

Figure 1.5 Linguistic processing model incorporating computational mind concepts.

and gustatory intensities that are acted upon by processes of generalization, deletion and distortion, mitigated by our internal perceptions, to become symbolic representations of our experience located in memory as the "mentalese" of deep structure (the "Storage" component in Fig. 1.5). In communicating our experience to others, our recall of deep structure progresses through the same cognitive processes of generalization, deletion and distortion from the symbolism of deep structure to surface structure linguistic and physiological behaviour that is personally, socially and culturally appropriate.

1.9 Conclusion

This first, and introductory, chapter brings together some specific ideas of research and practice that have significantly contributed to the three target areas of this book, *Networks in Society: Links and Language*. In doing so, some key perspectives emerge that enable general models or meta-representations of cognitive function as it relates to language, communication and social interaction and thus the formation and maintenance of social structure (networks) to be developed.

By way of introduction, the intent is to provide a launching point for other chapter authors as they expand on some of the introductory ideas or introduce other concepts to the mix. In addition, this chapter aims to stimulate debate around these topics in the broader community and to identify and propose various models that might act as focus points for researchers to pursue specific areas of interest. I hope that it will generate more questions than answers!

References

1. Anderson, J. A. (1995). *An Introduction to Neural Networks* (The MIT Press, Cambridge, MA).
2. Badham, J., and Stocker, R. (2010a). Impact of network clustering and assortativity on epidemic behaviour. *J. Theor. Population Biol.* **77**(1), pp. 71–75.
3. Badham, J., and Stocker, R. (2010b). A spatial approach to network generation on three properties: degree distribution, clustering coefficient and degree assortativity. *J. Artif. Soc. Social Simul.* **13**(1), p. 11.
4. Bandler, R., and Grinder, J. (1975). *The Structure of Magic*, Vol. 1 (Science and Behaviour Books, Palo Alto, CA).
5. Bandler, R., and Grinder, J. (1976). *The Structure of Magic*, Vol. 2 (Science and Behaviour Books, Palo Alto, CA).
6. Bandler, R., and McDonald, W. (1985). *Insider's Guide to Submodalities* (Meta Publications, Cupertino, CA).
7. Barabási, A.-L., and Bonabeau, E. (2003). Scale-free networks. *Sci. Am.* **288**, pp. 50–59.
8. Barak, A., and Weigt, M. (2000). On the properties of small-world network models. *Eur. Phys. J. B* **13**, pp. 547–560.
9. Berkowitz, S. D. (1982). *An Introduction to Structural Analysis* (Butterworths, Toronto).
10. Bodenhamer Bob G., and Hall, L. Michael. (2007). *The User's Manual for the Brain* Vol. I (Crown House Publishing Company LLC, Bethel, CT).
11. Burger, M. J., and Buskens, V. (2009). Social context and network formation: an experimental study. *Soc. Networks* **31**, pp. 63–75.

12. Buzan, T. (2010). *The Mind Map Book: Unlock Your Creativity, Boost Your Memory, Change your Life* (BBC Active, Pearson Education Group, Essex, UK).
13. Chomsky, N. (1957). *Syntactic Structures* (Mouton, The Hague, Netherlands).
14. Chomsky, N. (1968). *Language and Mind* (Harcourt Brace Jovanovich, Inc., New York, NY).
15. Churchland, P. S., and Sejnowski, T. J. (1993). *The Computational Brain* (The MIT Press, Cambridge, MA).
16. Damasio, A., and Damasio, H. (1992). Brain and language. *Sci. Am.* **267**, pp. 88–95.
17. Dilts, R., Grinder, J., Bandler, R., and Delozier, J. (1980). *NLP: The Study of the Structure of Subjective Experience*, Vol. 1 (Meta Publications, Capitola, CA).
18. Dilts, R. (1998). *Modelling with NLP* (Meta Publications, Capitola, CA).
19. Doreian, P. (1979). On the evolution of group and network structure. *J. Soc. Networks* **2**, pp. 235–252.
20. Dunbar, R. I. M. (1992). neocortex size as a constraint on group size in primates. *J. Human Evol.* **20**, pp. 469–493.
21. Dunbar, R. I. M. (1993). Co-evolution of neocortex size, group size and language in humans. *Behav. Brain Sci.* **16**(4), pp. 681–735.
22. Dunbar, R. I. M. (1995). Neocortex size and group size in primates: a test of the hypothesis. *J. Human Evol.* **28**, pp. 287–296.
23. Dunbar, R. (1996). *Grooming, Gossip and the Evolution of Language* (Faber, London).
24. Erdös, P., and Rényi, A. (1960). On the evolution of random graphs. *Mat. Kutato Int. Kozl.* **5**, pp. 17–61.
25. Gilbert, Nigel. (1997). Emergence in social simulation, in *Artificial Societies: The Computer Simulation of Social Life*, ed. Gilbert, N., and Conte, R. (University College London [UCL] Press, London), pp. 144–156.
26. Gilbert, N. (2002). Varieties of Emergence (ed. transcript). Agent 2002 Social Agents: Ecology, Exchange and Evolution Conference.
27. Gilbert, N., and Troitzsch, K. G. (2005). *Simulation for the Social Scientist* 2nd ed. (Open University Press, Buckingham, UK).
28. Gilbert, N. (2008). *Agent Based Models: Quantitative Applications in the Social Sciences* (Sage Publications, London, UK).

29. Gilbert, N., and Ahrweiler, P. (2009). The epistemologies of social simulation research, in *Epistemological Aspects of Computer Simulation in the Social Sciences: Second International Workshop*, EPOS 2006, Brescia, Italy, October 5–6, 2006, Revised Selected and Invited Papers (Springer-Verlag), pp. 12–28.
30. Gladwell, M. (1999). *The Tipping Point* (Little Brown, London).
31. Goldspink, C. (2002). Methodological implications of complex systems approaches to sociality: simulation as a foundation for knowledge. *J. Artif. Soc. Social Simul.* **5**(1) p. 3.
32. Green, D. G. (2002). Hierarchy, complexity and agent models, in *Our Fragile World* (UNESCO, Paris).
33. Hebb, D. (1949). *The Organisation of Behaviour: A Neuropsychological Theory* (Wiley, New York).
34. Holland, J. (1995). *Hidden Order: How Adaptation Builds Complexity* (Perseus Books, Reading, MA).
35. Holyst, J. A., Kacperski, K., and Schweitzer, F. (2000). Phase transitions in social impact models of opinion formation. Elsevier Preprint arXiv:cond-mat/0004026. **1**(1), pp. 1–21.
36. Kandel, E. R., and Hawkins, R. D. (1992). The biological basis of learning and individuality. *Sci. Am.* **267**(3), pp. 78–86.
37. Knoke, D., and Yang, S. (2008). *Social Network Analysis: Quantitative Applications in the Social Sciences*, 2nd ed. (Sage Publications, Los Angeles).
38. Koster, R. (2005). *A Theory of Fun for Game Design* (Paraglyph Press, Scottsdale, AZ).
39. Langton, C. G. (1990). Computation at the edge of chaos: phase transitions and emergent computation. *Physica D* **42**, pp. 12–37.
40. Lewis, B., and Pucelik, F. (1990). *Magic Demystified: A Pragmatic Guide to Communication and Change* (Metamorphous Press, Portland, OR).
41. Pinker, S. (1994). *The Language Instinct: How the Mind Creates Language,* (HarperCollins Publishers, New York).
42. Pinker, S. (1997). *How the Mind Works* (W. W. Norton, New York).
43. Pinker, S. (2007). *The Stuff of Thought: Language as a Window into Human Nature* (Penguin Books, London).
44. Scott, J. (2000). *Social Network Analysis: A Handbook* (Sage Publications, London).
45. Stocker, R., Cornforth D., and Bossomaier, T. R. J. (2002). Network structures and agreement in social network simulations. *J. Artif. Soc. Social Simul.* **5**(4), p. 3.

46. Stocker, R., Cornforth, D., and Green, D. G. (2003). A simulation of the impact of media on social cohesion. *J. Adv. Complex Sys.* **6**(3), pp. 349–359.
47. Wasserman, S., and Faust, K. (1995). *Social Network Analysis: Methods and Applications* (Cambridge University Press, Cambridge, MA).
48. Watts, D. J., and Strogatz, S. H. (1998). Collective dynamics of "small-world" networks. *Nature* **393**, p. 440.
49. Watts, D. J., Dodds, P. S., and Newman, M. E. J. (2002). Identity and search in social networks. *Science* **296**, pp. 1302–1305.
50. Wellman, B. (1988). Structural analysis: from method and metaphor to theory and substance, *Social Structures: A Network Approach*, ed. B. Wellman and S. D. Berkowitz (Cambridge University Press, Cambridge, MA), pp. 19–61.
51. Wellman, B., and Berkowitz, S. D. (1988). Introduction: studying social structures, in *Social Structures: A Network Approach*, ed. B. Wellman and S. D. Berkowitz (Cambridge University Press, Cambridge, MA), pp. 1–18.

Chapter 2

On Graphs, Networks and Social Groups

David G. Green[a] **and Terry Bossomaier**[b]
[a]*Centre for Intelligent Systems Research, Monash University, Wellington Road, Clayton, Victoria 3800, and*
[b]*Centre for Research in Complex Systems (CRiCS), Charles Sturt University, Bathurst, New South Wales 2640, Australia*
david.green@monash.edu, tbossomaier@csu.edu.au

Before examining networks in society, we must first understand the nature of networks themselves. Graphs and networks occur in many different contexts, so their properties underlie features that are seen in all the systems where they occur. In this chapter we introduce some of these properties and show how they manifest themselves in social systems. We also introduce some of the tools that can be used to study social networks.

2.1 Graphs and Networks

Put simply, a *graph* is a set of objects that are linked together in some way. In more formal terms, we refer to the objects as *nodes* or *vertices*. These are linked by *edges*, which join nodes together in pairs. We can describe many systems in these terms: towns are linked by roads; companies are linked by contracts; and people are

Networks in Society: Links and Language
Edited by Robert Stocker and Terry Bossomaier

ISBN 978-981-4316-28-6 (Hardcover), 978-981-4364-82-9 (eBook)
www.panstanford.com

linked by their relationships to one another. A graph in which the edges have direction is called a *directed graph*.

A *network* is simply a graph in which the nodes and/or edges have values associated with them. A family tree, for instance, is a network in which the nodes are people with names, and the edges are family relationships, such as "A is parent of B". In reality, there are attributes associated with the nodes or edges in virtually any graph. For this reason, we will use the terms "graph" and "network" interchangeably throughout this chapter, except where we are talking about properties of graphs specifically. Unless we specify the nature of a network, assume that we are referring to non-directed networks.

2.1.1 *Basic Terms and Concepts*

A number of terms are used routinely in discussing graphs and networks:

- If an edge in a graph links node A to node B, then we say that A and B are *connected*. Also they are *adjacent* to one another and A and B are said to be *neighbours* (Fig. 2.1).
- The *neighbourhood* of a node A consists of all nodes that are its neighbours.
- A *path* between a set of points A and B is a sequence of adjacent nodes, starting with A and ending with B, for instance A-T, T-U, U-V, V-B (Fig. 2.1).
- Two nodes in a network are said to be *connected* if there is a path joining them. A network is *connected* if every pair of its nodes has at least one path between them. The network is *fully connected* if every node is adjacent to every other node.
- The *length* of a path is the number of edges in it. The *distance* between two nodes is the length if the shortest path joining them (note that the distance is undefined if the nodes are not connected).
- The *diameter* of a network (also called the *characteristic path length*) is the maximum distance separating a pair of its nodes. For a disconnected graph, the diameter is not well defined. However, it is sometimes defined to be the maximum diameter of its connected subgraphs.

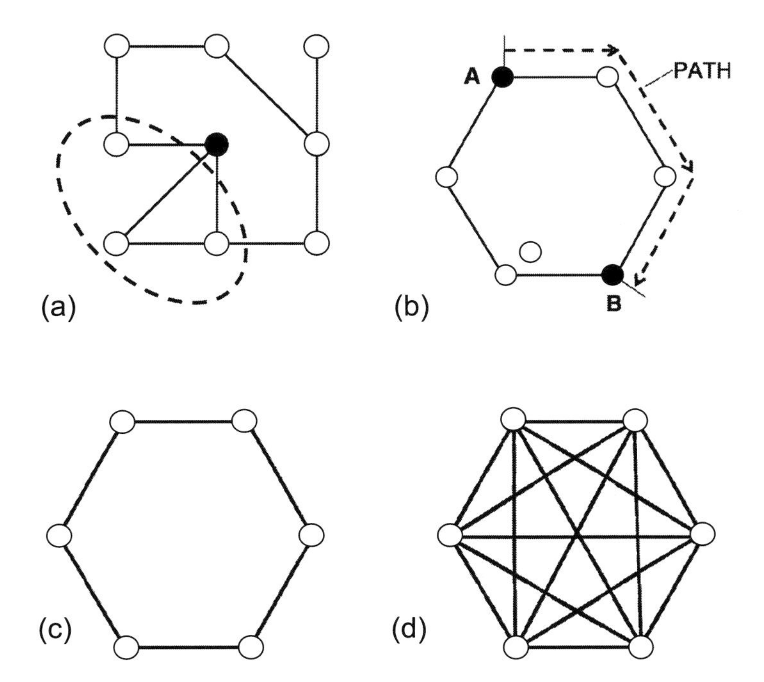

Figure 2.1 Some basic concepts about simple graphs and networks. Nodes are drawn as small circles, and edges as lines joining pairs of circles. (a) The nodes enclosed by the dotted loop are the neighbours of the node shaded black. (b) The dotted line shows a path from node A to node B. The diameter of this graph is 3. (c) A connected graph. With 6 nodes, there are 15 possible edges, so the edge density $\varepsilon = 6/15 = 0.4$. (d) A fully connected graph ($\varepsilon = 1.0$).

- In an undirected graph, the *degree* of a node is the number of edges linking it to other nodes. So the degree is also the number of nodes in its neighbourhood. For directed graphs, we need to distinguish between a node's *in-degree* (the number of edges leading to it) and its *out-degree* (the number of edges leading from it).
- The *edge density* ε of a graph (sometimes known as its *connectance* or *connectivity*) is the ratio between the number of edges in the graph and the maximum possible number of edges. For an undirected graph containing N nodes, there is a maximum of $N(N-1)/2$ edges (assuming

no edge links a node to itself), so if the graph contains E edges, the edge density ε is given by

$$\varepsilon = \frac{2E}{N(N-1)}. \tag{2.1}$$

For directed graphs, the maximum number of edges is $N(N-1)$, so the numerator above changes to E instead of $2E$.

- A graph is called *sparse* if its edge density is low.
- A *cluster* is a group of nodes that are well connected. Note that if a network's edge density is high, then it is likely to include clusters by chance.

2.2 Representing Networks

The traditional way to represent networks pictorially is as dots joined by lines (Fig. 2.2). In directed graphs the edges (also called *arcs*) are represented by arrows. For instance we might denote "A is a parent of B" by A→B. Edges can also be represented as pairs, such as (A,B).

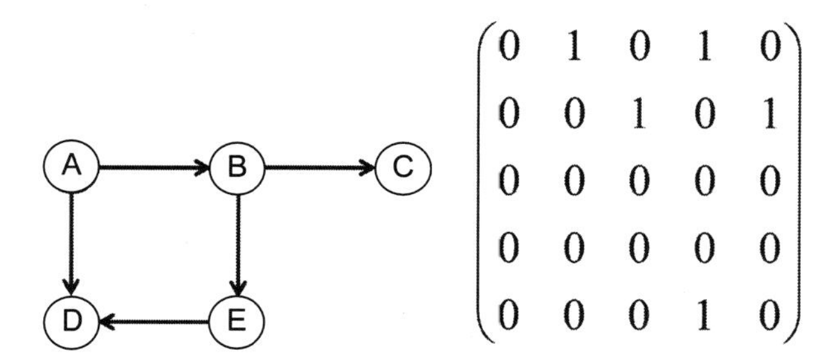

$$\begin{pmatrix} 0 & 1 & 0 & 1 & 0 \\ 0 & 0 & 1 & 0 & 1 \\ 0 & 0 & 0 & 0 & 0 \\ 0 & 0 & 0 & 0 & 0 \\ 0 & 0 & 0 & 1 & 0 \end{pmatrix}$$

Figure 2.2 Two ways of representing a simple network. At left is a simple network with five nodes (labeled A–E) and five arcs (directed edges). At right is the corresponding adjacency matrix. The rows and columns of the matrix correspond to the nodes A–E in order. Each non-zero entry denotes an arc from the node represented by its row to the node represented by its column. In networks where the edges are undirected, the equivalent adjacency matrix would be symmetric about the diagonal. That is, they would be entered in cells by swapping rows and columns. For instance, the entry of 1 in column 4, row 1 would be repeated in the cell at row 4 and column 1.

Simple networks, and even some fairly complex ones, can be represented conveniently as a diagram consisting of points and lines (Fig. 2.2). However, in text format, we normally need to list the vertices, and then list all the edges one by one. So a network with four nodes A, B, C, D might contain edges (A,B), (A,C), (A,D), (B,C), (B,D).

Another way to indicate which pairs of nodes are joined by edges is to use an *adjacency matrix*. This is a table in which the rows and columns each correspond to the nodes of the graph. A cell in the body of the table contains a 1 if there is an edge linking the row vertex to the column vertex, and a 0 otherwise (Fig. 2.2). Another matrix representation (not shown) is the *incidence matrix* I_{ij}, in which rows denote nodes and columns denote edges. An entry of 1 in a cell I_{ij} indicates that the node for row i is on endpoint of the edge for column j.

2.2.1 *Common Network Topologies*

Certain structures are commonly found in the patterns formed by edges in a graph. These are not all mutually exclusive. In some cases they are implicit in the way a social network is formed [1].

A *random network* is one in which edges are assigned at random to pairs of nodes. One way to test whether a graph is random is to check that the degrees $d(v)$ of the vertices v follow a Poisson distribution [2]. That is, the probability $p(d;\lambda)$ of a node having degree d is given by

$$p(d;\lambda) = \frac{\lambda e^{-\lambda}}{d!}, \qquad (2.2)$$

where λ is the expected degree. In a random, undirected graph with n vertices, λ is related to the edge density ε by

$$\lambda = \varepsilon(n-1). \qquad (2.3)$$

Random networks have some important properties, which we will look in the next section.

A *cycle* is a graph that consists of a single closed path (Fig. 2.3). That is, the path starts and ends at the same node, for example A-B-C-D-E-A. Cycles often occur in networks of dynamic interactions, where they lead to feedback.

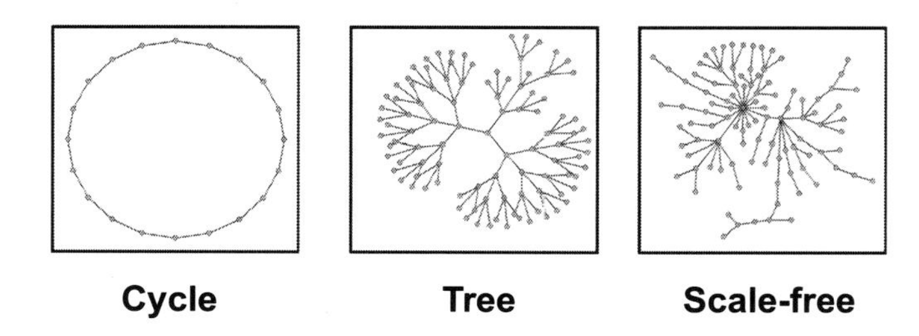

Figure 2.3 Examples of networks with some common topologies. All three are connected graphs. See text for further discussion.

A *tree* is a connected graph that contains no cycles (Fig. 2.3). Trees are closely associated with hierarchies. Large organizations, for instance, are often arranged hierarchically into divisions, departments, sections and so on. As structures, however, trees are brittle: breaking a single link splits the entire into two parts.

Beyond the global connectivity structures discussed above, trees, cycles and so on, there are a number of other properties that are useful in studying social networks. A few are presented below: modularity, motifs and mixing.

2.2.1.1 Modularity

Trees are closely related to the idea of modules, which are subgraphs that occur within a large graph. More precisely, a module is a subgraph that is well connected internally, but poorly connected to the rest of a graph. Modules are related to trees because they can be hierarchical in nature. That is, a module may be composed of smaller modules (see Fig. 2.4).

Many networks in the natural and man-made world have a *modular* structure. Consider a social network made up of school kids. We can see several different sorts of networks, such as the network of kids who play competitive sport or the network of kids playing in the school orchestra. Now some will play both football and the flute, but we might find that the musicians tend to hang out together more than they hang out with footballers.

The above very simple example shows how difficult the issue of modularity is. There is an immediate problem in how we define the

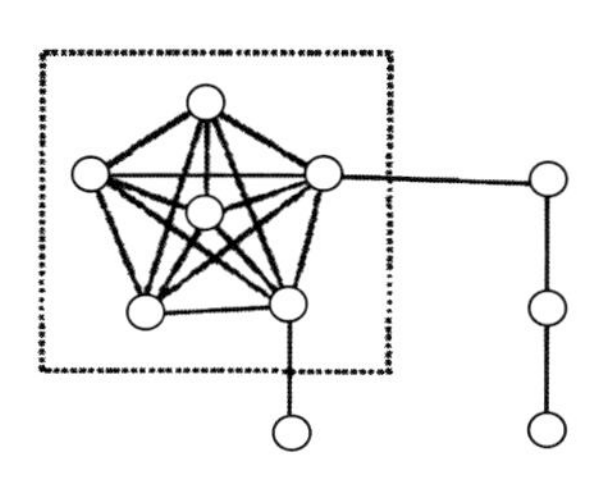

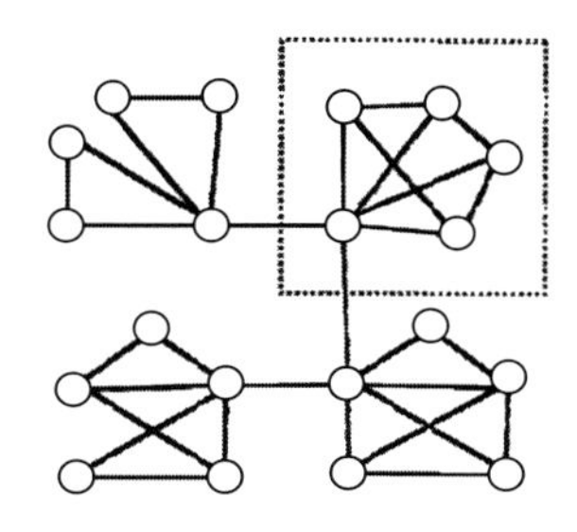

Figure 2.4 Examples of clusters and modules. The graph at left contains a single tight cluster (indicated by a box). At right is a modular graph (one module is indicated by a box). Each module is well-connected internally, but in each case, only a single edge links pairs of modules.

links. If we use a threshold amount of time hanging out together, we would get a different network structure if we included the time engaged within the activity, music or sport, as opposed to time spent outside school, say. Thus modules are not clear cut, and they could be defined in lots of different ways. Section 2.2.2 takes up the issue of definition in more detail.

Apart from the issue of defining modularity there is also the issue of how it arises. There are many folklore possibilities—"Birds of a feather flock together", and so on—but there have been some interesting theoretical insights which have appeared quite recently. Lipson et al. [3] proposed that changing environments lead to modular structures. The idea here is that it is easier to respond to environmental change with a modular toolkit, where one can just plug in different modules, or adapt specific modules to the change. Kashtan and Alon [4] showed that periodic switching between goals caused the evolution of modular network structures.

2.2.1.2 Motifs

Closely related to modules is the idea of motifs. A motif is a well-defined pattern of local connections within a network (Fig. 2.5). Usually they are relatively small and are associated with some particular aspect of the system concerned. In dynamic systems, for

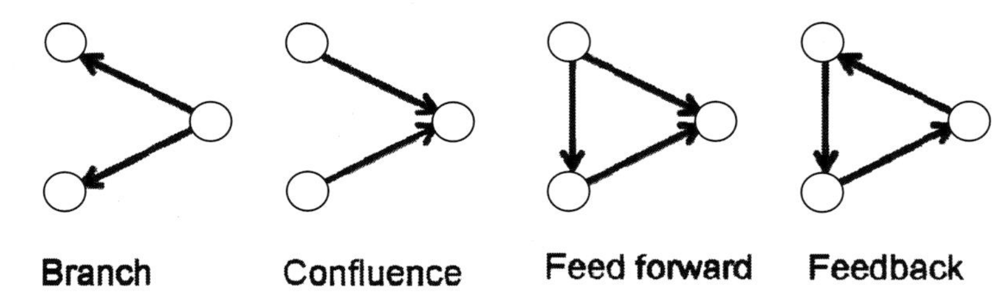

Figure 2.5 Some examples of 3 node motifs that occur within directed graphs.

instance, feedback loops are important motifs, with negative loops being important contributors to system stability and positive loops promoting instability. Some networks topologies, such as loops and trees, can be treated as motifs when they appear as elements within larger networks.

Motifs have figured in research on networks for many years. In bioinformatics the method of searching for motifs as indicators of structure and function became popular as a way of obtaining clues about protein folding patterns [5]. A review of motif searching by Milo et al. [6] triggered an upsurge interest in searching for motifs as a way of interpreting networks in general.

Modules and motifs are complementary ideas. In terms of organization within complex networks, they can be thought of as top-down versus bottom up approaches to creating order. Modules are often associated with ways of partitioning a large system into smaller, more manageable parts. Motifs, on the other hand, are usually identified as elements of local order. Note that the term "module" is also used to denote self-contained elements, often in the sense that they can be replicated (especially in engineering). When treated this way, motifs and modules may refer to the same things.

2.2.1.3 Mixing

Another network property that is quite a strong characteristic of social networks is assortativity or its opposite, disassortativity. Imagine that in the school music network the singers are the most popular, soccer players are most popular in sport and singers and soccer players do not mix very much. We have here an example of disassortative mixing [7, 8]. The highly connected singers are not connected to the other highly connected kids, the soccer players. Assortativity is the opposite. Highly connected nodes (kids) in the network are most likely to be directly connected to other highly connected nodes. One can imagine this occurring with physical attractiveness. The most attractive people tend to make the most socially desirable acquaintances and they have lots of links. They might also tend to hang out together, making the highly connected nodes directly connected to other highly connected nodes.

Mark Newman proposed that social networks are most likely to be assortative and found several powerful examples. But the story is not clear cut and other networks have now been found that are disassortative. Other structures have also been advocated, such as core-periphery networks, wherein there is a tightly connected core with a loosely connected periphery [9].

Scale-free networks are connected networks in which the degree distribution of the vertices follows a power law [10]. This means that a few nodes are highly connected, but most have just one or two connections (Fig. 2.3). More precisely, the probability of a node having degree d is given by

$$p(d) = ad^{-k}, \tag{2.4}$$

where a and k are constants. Networks of this type usually form when nodes are added progressively to the network and form edges to existing nodes by preferential attachment. That is, they tend to form links to nodes that already have many links to other nodes.

Small worlds are connected networks that are sparse, highly clustered, and yet have low diameter [11]. They typically appear when a network consists chiefly of "local" connections, but also contains a smattering of "local-range" connections (Fig. 2.6). Social networks are typically small worlds [12, 13]. In a business, for example, everyone knows the people they work with every day, but they also have some more distant associates, in other organizations. These long-range connections reduce the overall diameter of the network.

Strictly speaking, the concept of small world is a property that graphs possess to a greater or lesser degree. Small worlds stand between regular graphs at the one extreme and random graphs at

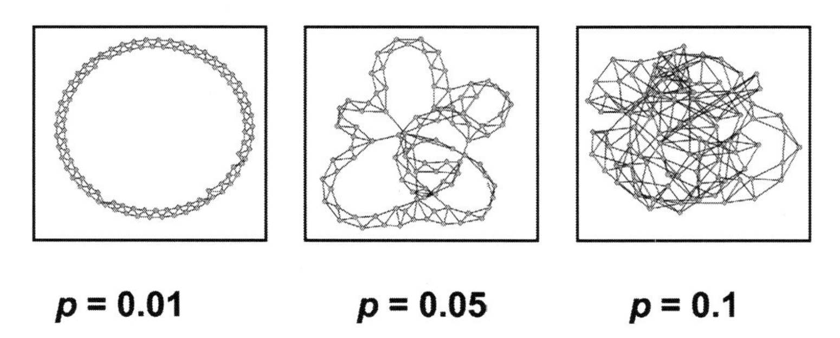

Figure 2.6 The small world property in networks. Starting from a regular network (left), edges are added to link randomly selected pairs of nodes. The small world effect can be seen clearly in the central network: edges have joined nodes that were previously on opposite sides of the network, thus reducing the network diameter considerably. As further edges are added (right) the network becomes indistinguishable from a random network.

the other (Fig. 2.6). One way to understand them is to imagine a regular graph, a cycle for instance. This regular network (of say N nodes) has a diameter of $N/2$. If we progressively add edges between random pairs of nodes, then these new edges create shortcuts between different parts of the cycle, thus reducing the overall diameter. However, they also randomize the overall structure until eventually the graph is indistinguishable from one that was formed at random.

The idea of small world inspired the popular term *six degrees of separation.* The idea is that no more than six steps are required to link any pair of people anywhere in the world. Perhaps the most famous example of a small world network is the network of relations between movie actors [14]. Films define the edges of this network. Two actors have a link (1 degree of separation) if they have appeared in the same film together [15]. Now one would expect that there would be considerable clustering between groups of actors working at the same time and for the same studios.

2.2.2 *Measures of Network Properties*

Several metrics have been devised to provide indicators of structural properties in networks. In the previous section we have already met some of these measures. One is the edge density, which indicates how richly connected the network is. A social network that is rich in connections has a high level of individual interactions. On the other hand, one that is poorly connected is likely to contain isolated individuals or cliques. The other measure described previously is the network diameter (characteristic path length). In a social network, this measure indicates how closely integrated the society is. We will see further implications associated with edge density and diameter in later sections.

The *clustering coefficient* indicates the extent of connectivity within local neighborhoods. For an individual node, the clustering coefficient is based on the number of connections between its immediate neighbours. If a node has K neighbours (in an undirected graph), then the maximum possible number of connections between those neighbours is $1/2K(K-1)$. So if there are (say) J edges between the node's neighbours, the clustering coefficient $C(v)$ is the ratio

$$C(v) = \frac{2J}{K(K-1)}. \tag{2.5}$$

The value of this formula is actually undefined if v has no neighbours at all, so in such a case the value is defined to be 0. Otherwise its value will range from 0 (if none of the neighbours are linked) to 1 (if the neighbourhood is fully connected.

The clustering coefficient $C(G)$ for an entire graph G is the average of the above over all nodes in the network. That is, for a graph G with N nodes,

$$C(G) = \frac{1}{N}\sum_{i=1}^{N} C(v_i). \tag{2.6}$$

As for individual nodes, the value of $C(G)$ ranges between 0 and 1. High values indicate the presence of local clustering within a graph. Note, however, that one problem with the measure is that it tends to be correlated with the edge density. That is, networks with high edge density necessarily have high values for $C(G)$.

Measures of *modularity* indicate the extent to which a network separates into distinct modules. It often happens that a network contains sets of nodes that we expect to form modules. In this case, we can test how strongly modular the expected structure is.

Suppose that a network of N node has E edges and let A_{ij} denote the adjacency matrix of the network. Now suppose that we partition the network by grouping the nodes into a set of subsets $C = \{c_i, i = 1 \ldots p\}$, such that every node belongs to exactly one subset c_i and every subset contains at least one node. If we take any two disjoint subsets c_i and c_j, then we can denote by e_{ij} the fraction of all edges that provide links between pairs of nodes in the two subsets. Finally, let a_i denote the fraction of all edges in the network that have nodes from c_i as one of their ends. That is,

$$a_i = \sum_{j=1}^{p} e_{ij}. \tag{2.7}$$

Using the above definitions, Newman and his colleagues introduced a measure of modularity Q, given by

$$Q = \sum_{i=1}^{p} (e_{ii} - a_i^2). \tag{2.8}$$

Values of Q range from 0 (no modularity) to 1 (heavily modular).

The Newman measure is quite successful, but Rosvall and Bergstrom [16] showed that it can be sensitive to the total number of links in the network. They propose an information theoretic framework, in which they search for the modularity structure that provides the most information about the actual network. In other words, the mutual information between the network and the modular representation is maximal. Suppose that in our network of kids who play sport, all the n_f soccer (football) players talk to each other, all the n_r rugby players talk to each other and all the n_c cricket players talk to each other. But everybody plays only one sport and only the captains talk to each other.

Suppose in total we have N kids (i.e., $N = n_c + n_f + n_r$). The number of possible networks is huge, 2^N, so the entropy, H, of the set of networks is $H = \log_2 2^N = N$ if they are all equally likely. If we know about this modular structure, then all we need to specify is the captains. The number of possible captain combinations between footballers and cricketers is $n_c n_f$, making the total number of possible links $n_c^2 n_f^2 n_r^2$, giving $H = 2\log_2 n_c n_f n_r$. If the number playing each sport is the same, $N/3$, then $H = 2\log_2 N/3$, *which* is *much* smaller.

So, in the case where the number of modules is known, Rosvall and Bergstrom minimize

$$H = \prod_i \binom{n_i(n_i - 1)}{l_{ii}} \prod_{i>j} \binom{n_i n_j}{l_{ij}}, \tag{2.9}$$

where n_i is the number of nodes in module i and l_{ii} is the number of edges in module i and l_{ij} is the number of links between modules i and j.

If the number of modules is not known it gets more complicated. This measure improves on Girvan and Newman for some networks, but is still not the final word. At the time of writing there is no unique or consensus measure for network modularity.

2.2.3 *Markov Random Graphs*

If we have a set of data obtained empirically then we might want to ask how well a given graph model will fit the data. We have already seen tests for assessing whether or not a graph is random. The idea of *Markov random graphs* (MRGs), takes us a step further [17].

If we have an adjacency matrix, A, and we observe a set of edges, b_{ij} on a given node set, then the probability that this graph would occur as an Erdös-Renyi graph (see next section) is

$$P(A = b) = \exp(\varepsilon L(b))/Z, \tag{2.10}$$

where ε is the edge density and $L(b)$ the number of edges and Z is a normalising constant.

Random graphs do not exhibit very many of the motifs referred to above, which are common in many social networks. To get around this, the *exponential random graphs* (ERGs) of Pattison and colleagues [18] introduce the idea of ties and configurations. A configuration is a motif, a (usually) small group of nodes with a particular connectivity structure. Examples are a 2-star (3 nodes with one connected to each of the others) and a triangle. Equation 2.1 now generalizes to incorporate these motifs:

$$P(A = b) = \exp(\varepsilon L(b) + \sigma_2 S_2(b) + \tau T(b))/Z, \tag{2.11}$$

where $S_2(b)$ and $T(b)$ are the number of 2-stars and triangles respectively and σ_2 and τ are control parameters [19]. Thus we get

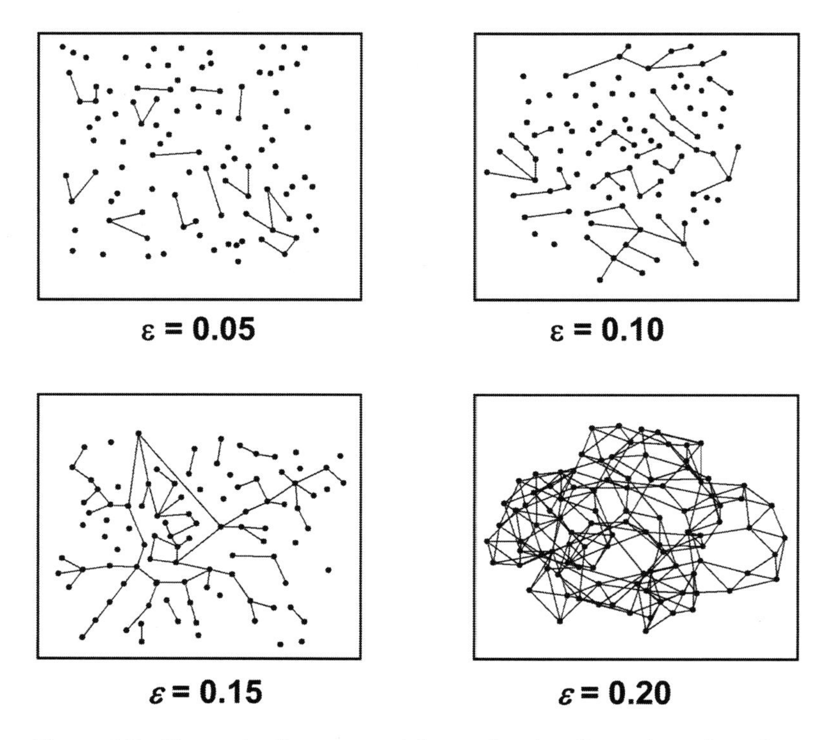

Figure 2.7 Stages in the connectivity avalanche. Examples of random networks with increasing edge density. For the 100 nodes (as shown here), the critical density is $\varepsilon = 0.01$. These networks show different stages of the connectivity avalanche and the formation of a unique giant component.

random graphs but with a differing density of particular motifs than would occur in the Erdös-Renyi process.

It is now possible to generalize Equation 2.2 to graphs which have arbitrary numbers of different configurations, C:

$$P(A = b) = \exp\left(\sum_C \eta_C g_C(b)\right) / Z, \tag{2.12}$$

where (η_C) is the parameter corresponding to configuration C and $g_C(b)$ is the number of of configurations of type C for Markov random graphs but can be generalized to other forms of network statistic [19]. Such models have been successfully applied to various canonical social network data sets such as the business networks of the Padgett Florentine families of medieval Florence.

2.2.4 *The Connectivity Avalanche*

In random networks the property with the most far-reaching consequences is the *connectivity avalanche*. Erdos and Renyi [20] found that when edges are added at random to a set of nodes the clusters formed do not grow in a regular fashion. Rather they form isolated small clusters (mostly pairs). When a critical density is reached, these small clusters rapidly merge to form a single, *unique giant component*, which absorbs all the remaining nodes as further edges are added (Fig. 2.7). In a network of N nodes, the connectivity avalanche occurs when the number of edges is $N/2$ (Fig. 2.8).

In essence the connectivity avalanche amounts to a phase change in connectivity. That avalanche transforms a network of many disconnected elements into a single connected network.

The connectivity avalanche underlies many critical phenomena [21]. Haken [22] argued that every critical phenomenon was associated with an *order parameter*, and the phase change occurred at some critical value of this parameter. For instance, temperature is the order parameter for water, which freezes at 0°C. This order parameter corresponds to the edge density of the underlying

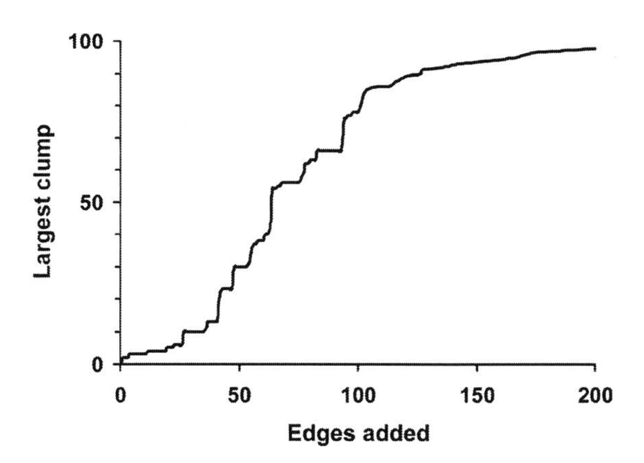

Figure 2.8 An experimental demonstration of the connectivity avalanche. Edges are added to randomly selected pairs of nodes in a set of 100 nodes (cf. previous figure). The graph shows the number of nodes in the largest clump of nodes. For this network the critical point occurs when 50 edges have been added.

network and the critical point is value at which the connectivity avalanche occurs [21].

A good example is an epidemic. The spread of a disease involves the transmission of a disease from one person to another. So an epidemic forms a network of infection in which people are the nodes and infections define the edges. In this case the infection rate defines the probability of an edge forming and substitutes for the edge density. If the infection rate is less than the critical threshold, then the epidemic stalls and only small numbers of people are infected. If the infection rate is high, then a large proportion of the population is approximately equal to the infection rate.

We shall explore some of the consequences of this property in the next section.

2.3 Networks and Complexity

2.3.1 *The Network Basis*

Networks are fundamental to the study of complex phenomena. Their importance stems from their universality. It can be shown that networks underlie the structure and dynamics of any complex system [21, 23]. This is usually clear since such systems often consist of many individual elements or agents that interact with one another. In a social community, for instance, people are the nodes and their relationships or interactions define the edges of the network.

Understanding the relevance of networks to behavioural dynamics is slightly less intuitive. Imagine a game of chess. The position of all the pieces at any given time defines the state of the game at that point. Each move changes the position of one or more pieces: it changes the state of the game. So each state of the game can be seen as a node of a network in which possible moves define the edges.

The above network model has several important implications. First, it means that properties of networks explain many common phenomena. The connectivity avalanche, which we saw earlier, is responsible for many features of complex systems. An epidemic, for instance, cannot spread if the density of edges in a behavioral

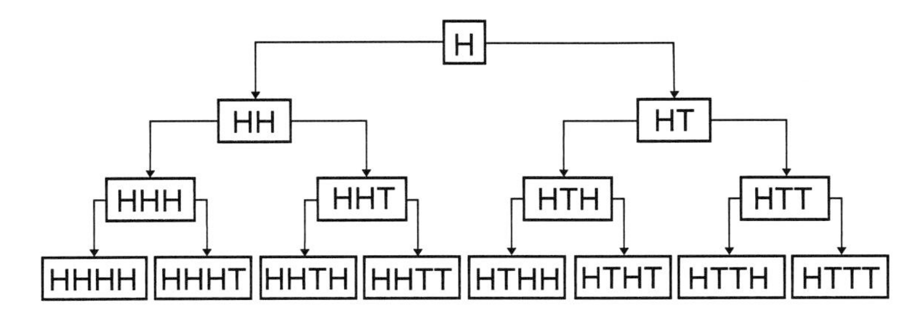

Figure 2.9 Part of the network of outcomes of a simple coin-tossing experiment.

network (i.e., the contacts between people) falls below the critical threshold.

Combinatorial complexity is intimately associated with networks. Take a simple case, such as tossing a coin several times. We can regard this as a network in which the nodes are sequences of results and each toss creates an edge linking one sequence to a longer sequence in which the next result is added (Fig. 2.9). For instance, one pathway (up to 6 coin tosses) might be as follows:

$$\mathrm{H} \rightarrow \mathrm{HH} \rightarrow \mathrm{HHT} \rightarrow \mathrm{HHTH} \rightarrow \mathrm{HHTHH} \rightarrow \mathrm{HHTHHH}$$

However, the network gathers complexity as it goes, because there are two possible branches at every stage (Fig. 2.9).

2.3.2 *Network Processes*

The most common approach to dealing with complexity is "divide and rule": that is, you break down a large complex problem into smaller, simpler parts. Companies, armies and other large organizations typically use this divide and rule to help manage their inherent complexity. An army, for instance, might be divided into divisions and these are composed of brigades or battalions, which are in turn composed of companies, which are then divided into platoons and sections. The effect of dividing large organizations this way is to eliminate complexities, such as two colonels giving conflicting orders to the same captain.

Several processes on networks are important sources of self-organization in complex systems. The first is feedback. Feedback occurs when the output of a process returns to become an input to it. Negative feedback tends to stabilize a system, preventing deviations from developing. Positive feedback, in contrast, destabilizes a system. For this reason, it is seen as something to be avoided at all costs in control systems. However, in complex systems, positive feedback is an important mechanism driving of self-organization. This is because it allows local events and differences to grow into system-wide features and patterns (see Chapter 4).

The connectivity avalanche, which we saw in the previous section, leads to two other common processes involved in self-organization. One is *self-organized criticality* (SOC) [24, 25]. This occurs where a system is constrained to remain in the critical region between connected and disconnected. In this situation, local disturbances create small connectivity avalanches, and the size of the avalanches follows a characteristic inverse power law. The classic example is a sand pile, in which addition of new grains of sand set off local collapses; a few collapses will be large, but most are small [25].

Another process arising from the connectivity avalanche is *dual-phase evolution* (DPE). This occurs in systems that normally reside in one phase (connected or disconnected) but are periodically pushed into the opposite phase by a disturbance of some kind [26–28]. Alternatively daily, monthly, annual or some other regular cycles of changes may be responsible for the phase changes. The crucial point is that different processes predominate in each phase. Usually these processes involve selection and variation in some form or other. A social example would be the switch between working in an office, where people interact only with immediate colleagues, versus attending a convention, where people interact with a wide range of people. New links are formed in the global phase and weak links are broken in the local phase. Some of the effects are discussed in Chapter 4.

2.4 Simulating Social Networks

In technical terms, the problem of simulating social networks resolves itself into two issues: representation and processing. Here we shall ignore simulations that model processes at large scale and consider only models that represent individual people and individual events (i.e., nodes and edges). Several paradigms have become widely used in modelling social systems.

Conceptually the simplest social models are Boolean networks. In these models, each person is represented by a node that has a single binary state variable as an attribute. Arcs (directed edges) denote peer influence (see [29]). Variations on this model have been used to investigate the formation of social consensus (e.g., [30]) and the evolution of cooperation (e.g., [31]) (for further discussion see Chapter 4). Variations on these networks include a richer set of attributes for each node.

Another important issue is the nature of the connections in the network. If an edge connects two individuals in a social network, what does that actually mean? If the edge denotes, say, family relationships, then it does not indicate an interaction, but rather a pathway that makes interactions possible or probable. Such connections are more or less permanent, as distinct from the intermittent connections that occur when individuals are close to one another in a crowd or other dynamic situation.

In computer science applications, a common paradigm is the BDI model for intelligent agents. This model, based on models of human reasoning [32], attributes individual agents with *beliefs*, *desires* and *intentions* that govern their actions [33]. However, such models have had only limited application in modelling social networks, chiefly in applications relating to advanced software and communications.

A more commonly used paradigm is *adaptive agents*. Some of the earliest studies were those by Hogeweg and Hesper, who introduced the idea of individual-based models to biology [34]. Besides embodying rules of behaviour for each individual, these models also incorporated a wide variety of "creatures" to facilitate running of the model. One set would prompt agents to act when special conditions occurred (e.g., when a lion should chase a nearby gazelle). Other observers would gather information about recurring patterns, which provided insights about the system concerned, as well as providing dynamic caching to accelerate processing. However, implementing these features required the development of considerable software machinery, which is not easy to duplicate.

A common problem in social modelling is dealing with spatially distributed agents. Two approaches are commonly used. One is to model the landscape as a grid of cells, with each cell representing an area of land (usually a square). Agents are identified by the cell they currently occupy. If the agents do not move about then they population may be represented by the cells themselves, otherwise the cell location is an attribute of the agent. Such a grid representation has the computational advantage of making it easy to identify spatial neighbours and to deal with local interactions. They also make it easy to incorporate monitoring data (especially satellite imagery) directly into a model. In doing so, grid models often draw on the paradigm of cellular automata [35–37]. CA models have been widely used to model epidemics [38] and other phenomena that involve spatial interactions.

The alternative approach is to indicate the location of each individual via exact coordinates. This representation has the advantage of locating an individual precisely within the landscape, thus avoiding artifacts that can arise from round-off errors associated with grid representations of space. On the other hand, they add computational costs through the need to identify changing patterns of neighbours.

In most simulation models, the states of individuals are updated synchronously—that is, in parallel. In social networks, these models assume that updating occurs in discrete time steps. Each step might represent a single day's activity, for example. However, if a model aims to represent individual encounters, then the outcome of one encounter could affect subsequent encounters. In such a case, a model would need to compute encounters asynchronously, in the order in which they occur.

A final issue to note about social simulation is that it can be difficult to model a particular system faithfully. Any model is perforce a simplification of reality, so it is virtually impossible to capture all the factors that motivate a person's actions at any given time and place. Similarly, a system may have hundreds of variables, so it may be impossible to calibrate them all to a sufficient degree of accuracy. In simplifying models, we try to separate the wheat from the chaff. That is we try to distill individual behaviour relative to some narrow domain, avoiding the complications posed by the rich variety of real life.

Finally, social networks are complex and complex systems of any kind are often chaotic, making them inherently unpredictable. This means that we cannot expect to use them in the same way as formula-based models of physical systems. Instead of trying to forecast the exact behaviour of the system under specific conditions, we need to test scenarios (e.g., worst cases) to explore what range of behaviour might be expected under particular conditions of interest. Likewise, we cannot explore a model's behaviour analytically as physicists might do with equations of motion. Instead, the modeller can use sensitivity analysis to explore how the system's behaviour might change in response to changes in particular conditions. This involves systematically running a model repeatedly, each time varying the value of a parameter or variable of interest, while keeping every other variable constant.

References

1. Newman, M. E. J., and Girvan, M. (2004). Finding and evaluating community structure in networks. *Phys. Rev. E* **69**, 026113 (15pp.).
2. Newman, M. E. J., Strogatz, S. H., and Watts, D. J. (2001). Random graphs with arbitrary degree distributions and their applications. *Phys. Rev. E* **64**, 026118 (17pp.).
3. Lipson, H. Pollack, J. B., and Sue, N. P. (2002). On the origin of modular variation. *Evolution* **56**(8), pp. 1549–1556.
4. Kashtan, N., and Alon, U. (2005). Spontaneous evolution of modularity and network motifs. *Proc. Natl. Acad. Sci. USA* **102**, pp. 13773–13778.
5. Bairoch, A. (1993). The PROSITE dictionary of sites and patterns in proteins. *Nucleic Acids Res.* **21**(13), pp. 3097–3103.
6. Milo, R., Shen-Orr, S., Itzkovitz, S., Kashtan, N., Chklovskii, D., Alon, U. (2002). Network motifs: simple building blocks of complex networks. *Science* **298**(5594), pp. 824–827.
7. Newman, M. E. J. (2002). Assortative mixing in networks. *Phys. Rev. Lett.* **89**(20), 208701.
8. Newman, M. E. J. (2003). Mixing patterns in networks. *Phys. Rev. E* **67**, 026126.

9. Jackson, M. O. (2008). *Social and Economic Networks* (Princeton University Press).
10. Barabási, A.-L., and Albert, R. (1999). Emergence of scaling in random networks. *Science* **286**(5439), pp. 509–512.
11. Watts D. J., and S. H. (1998). Strogatz, collective dynamics of "small-world" networks. *Nature* **393**, pp. 440–442.
12. Farkas, I., Derenyi, I., Jeong, H., Neda, Z., Oltvai, Z. N., Ravasz, E., Schubert, A., Barabasi, A.-L., and Vicsek, T. (2002). Networks in life: scaling properties and eigenvalue spectra. *Physica A* **314**, pp. 25–34.
13. Milgram, S. (1967). The small-world problem. *Psychology Today* **1**, pp. 60–67.
14. Collins, J. J., and Chow, C. C. (1998). It's a small world. *Nature* **393**, pp. 409–410.
15. The Oracle of Bacon, http://oracleofbacon.org/
16. Rosvall, M., and Bergstrom, C. T. (2007). An information-theoretic framework for resolving community structure in complex networks. *Proc. Natl. Acad. Sci. USA* **104**, pp. 7327–7331.
17. Frank, O., and Strauss, D. (1986). Markov graphs. *J. Amer. Stat. Assoc.* **81**, pp. 832–842.
18. Pattison, P. E., and Robbins, G. L. (2002). Neighbourhood-based models for social networks. *Sociol. Methodol.* **32**, pp. 301–337.
19. Robbins, G. L., Snijders, T., Wang, P., Handcock, M., and Pattison, P. E. (2007). Recent developments in exponential random graph (p^*) models for social networks. *Social Networks* **29**, pp. 192–215.
20. Erdos, P., and Renyi, A. (1960). On the evolution of random graphs. *Mat. Kutato. Int. Kozl.* **5**, pp. 17–61.
21. Green, D. G. (2000). Self-organization in complex systems. In T. R. J. Bossomaier and D. G. Green, eds. *Complex Systems* (Cambridge University Press), pp. 7–41.
22. Haken, H. (1981). *The Science of Structure: Synergetics* (Van Nostrand Reinhold, New York).
23. Green, D. G. (1993). Emergent behaviour in biological systems. *Complexity International* **1**. www.csu.edu.au/ci/vol1/David.Green/paper.html
24. Bak, P., Tang, C., and Weisenfeld, K. (1987). Self-organized criticality: an explanation of 1/f noise. *Phys. Rev. Lett.* **59**, pp. 381–384.
25. Bak, P. (1999). *How Nature Works: The Science of Self-Organized Criticality* (Springer-Verlag Telos). Reprint edition.

26. Green, D. G., Leishman, T. G., and Sadedin, S. (2006). Dual phase evolution: a mechanism for self-organization in complex systems. *InterJournal* ISSN 1081-0625, pp. 1–8.
27. Liu, J., Abbass, H., Green, D. G. (2011). *Dual Phase Evolution: From Theory to Practice* (Springer, Berlin).
28. Paperin, G., Green, D. G., Sadedin, S. (2011). Dual phase evolution in complex adaptive systems. *J. Royal Soc. Interface*, doi: 10.1098/rsif.2010.0719.
29. For some simple demonstrations see the VLAB social simulations: vlab.infotech.monash.edu.au/simulations/networks/social-network/
30. Stocker, R. Green, D. G., and Newth, D. (2001). Consensus and cohesion in simulated social networks. *J. Artif. Soc. Soc. Simul.* **4**(4), http://www.soc.surrey.ac.uk/JASSS/4/4/5.html
31. Abramson, G., and Kuperman, M. (2001). Social games in a social network, *Physical Review E* **63**, 030901.
32. Bratman, M. E. (1987). *Intention, Plans, and Practical Reason* (Harvard University Press, Cambridge, MA).
33. Rao, A. S., and Georgeff, M. P. (1991). Modeling rational agents within a BDI-architecture, in *Proceedings of the Second International Conference on Principles of Knowledge Represdentation and Reasoning*, ed. Allen, J., Fikes, R., and Sandewall, E. (Morgan Kauffman, San Mateo, CA), pp. 473–484.
34. Hogeweg, P., and Hesper, B. (1983). The ontogeny of the interaction structure in bumblebee colonies: a MIRROR model. *Behav. Ecol. Sociobiol.* **12**, pp. 271–283.
35. Hogeweg, P. (1988). Cellular automata as a paradigm for ecological modeling, *Appl. Math. Comput.* **27**, pp. 81–100.
36. Green, D. G. (1997). Modelling plants in landscapes, in *Plants to Ecosystems: Advances in Computational Life Sciences*, M. Michalewicz, ed. (CSIRO Division of Information Technology, Melbourne), pp. 85–96.
37. Green, D. G., Klomp, N. I., Rimmington, G. R., and Sadedin, S. (2006). *Complexity in Landscape Ecology* (Springer, Amsterdam). http://www.csse.monash.edu.au/~dgreen/books/kluwer/
38. For a simple demonstration see the VLAB epidemic simulation: vlab.infotech.monash.edu.au/simulations/cellular-automata/epidemic/

Chapter 3

Language Networks

Terry Bossomaier
Centre for Research in Complex Systems (CRiCS),
Charles Sturt University, Bathurst, New South Wales 2640, Australia

3.1 Introduction and Overview

Language is one of the defining features of human society. This chapter takes a look at the importance of language in social networks [4] and the role it plays in socioeconomic agent models. Chapter 2 describes some of the software tools useful in agent modelling applications.

Languages are not static, but dynamically shifting and evolving, and unfortunately many are going extinct. With currently around 6000 languages in the world, they are disappearing at an alarming rate. David Crystal [17] reports 51 languages with only one speaker in his 2000 book *Language Death*, so some of these will have died with their last remaining speaker as did Ubuh on October 8, 1992, and Kasabe on November 4, 1995.

At the time of writing, Surel Mofu at Oxford University is recording the last three speakers of the Indonesian language (Dusner [55]). Fischer [29] suggests that in a decade or so there

Networks in Society: Links and Language
Edited by Robert Stocker and Terry Bossomaier

ISBN 978-981-4316-28-6 (Hardcover), 978-981-4364-82-9 (eBook)
www.panstanford.com

may be just a handful of languages remaining, of which English will certainly be one, given its dominance on the Internet.

Perhaps surprisingly it took some time for the evolution of language to become accepted. Kenneally [39] describes how when Pinker and Bloom first proposed an evolutionary model, there was major opposition. But now great progress has been made in understanding language change over time. The details are beyond the scope of this chapter, but the interested reader will find Guy Deutchser's book *The Unfolding of Language* [19] engrossing and informative.

On a very long time scale, languages appear and disappear, but any given language, such as English, is continually changing, and developing dialects. It is this differentiating of social groups through language variations which is one of the most interesting features relevant to a modelling perspective.

In English the variations are profound across a relatively tiny geographic area. A century ago, the language of the BBC was a distinctly King's (or Queen's) English. Regional accents were rarely heard. That has of course changed, but the discrimination against language dialects and subcultures has also played out the other way around with language, specifically identifying cultures.

Teenagers frequently adopt their own slang to differentiate their peer groups from their parents. But then sometimes what is cool to them becomes cool to everybody. The case of cockney rhyming slang may have had a similar agenda, specifically to hide information. Porkies (*porkie, pork pie, lie*) were a necessary tool of the London underworld.

Languages change in several ways. The grammar changes (most of the case endings of Latin have disappeared in modern-day Italian), words change their meaning and new words appear. Although grammar does distinguish language subgroups, the changes in semantics, the meanings of words, are usually more important.

But grammar and words are tightly coupled, with words splitting or fusing in different ways. In English we have a *firework display.* In German there is just the one word, *Feuerwerkveranstaltung.* This is not that different from the English (*Feuerwerk,* firework; *Veranstaltung,* display), but the words have been joined together. *Veranstaltung* is itself an assemblage of *ver, an* and *staltung.* In

French we have all these pieces split apart in *spectable de fusée de feu d'artifice* (see Section 3.2.1).

The sounds of language are of course immensely important, and much has been written about their evolution and spatial distributions. But our focus here is much more on aspects of the concept and cultural representation within language, and thus we will not pursue phonetics in any detail other than for a brief look at phonological networks (Section 3.3.6).

3.2 A Network Perspective on Language

The study of language goes back a long way, to at least Aristotle, but the discipline of linguistics really took off in the latter part of the 19th century, driven in part by the seminal Swiss linguist Ferdinand de Saussure. Linguistics has seen several paradigm shifts in its short history, such as Chomsky's formalizing of grammatical structure and introduction of the idea of a universal grammar [13]. In fact it may be going through another transition at the present time, driven not by studying human cultures and societies, but by the explosion of information on the World Wide Web.

Searching for things on the web began slowly with numerous pretty search engines, and then Google came along and the world changed. Adding a new word to English (to *google*), and adding a new concept to every language of the developed world, Google rose to prominence through the invention of a clever new algorithm for ranking web pages according to how well they would match a text-based search query.

Thus being able to search out relevant pages, *without any understanding of content*, stimulated a great deal of new work in how to build semantic relationships between words and use them for search. More or less at the same time, graph theory, a fairly difficult and obscure branch of mathematics, suddenly became a rapid growth area in the study of biological, social, and computational networks, as discussed in Chapter 2. At the same time as these computing and theoretical developments were taking place, rapid advances were also occurring in imaging of the human brain while it is carrying out various sorts of tasks, including conceptual and

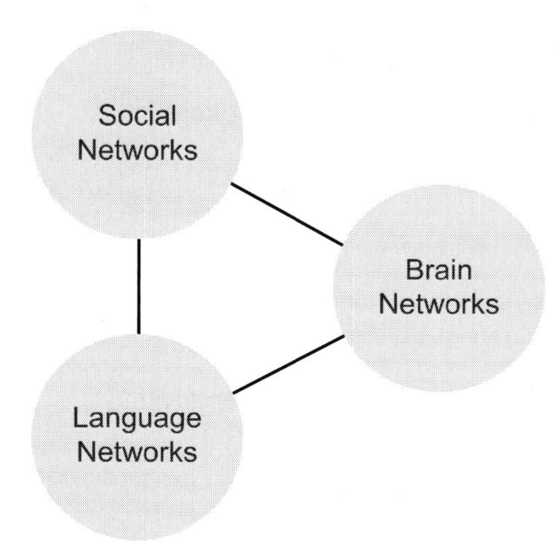

Figure 3.1 Social, language and brain networks.

linguistic tasks. Starting perhaps a little later was the surge of interest in social media networks, dominated at present by Facebook and Twitter.

Thus, this chapter provides some background to understanding these network dimensions, shown in Fig. 3.1. Studies vary from overarching perspectives on complex networks and language [38] to fine-grained studies of grammatical and semantic properties.

3.2.1 *Language Families*

One of the major achievements of linguistics has been the cataloguing of languages, which have a similar taxonomic structure to plants and animals. Languages may be grouped into families and superfamilies. Most European languages (Hungarian and Finnish being notable exceptions) fall into the superfamily Indo-European. Nobody actually speaks Indo-European, but painstaking scholarship has inferred this historical parent of dozens of languages, alive and extinct. Within this superfamily there are 10 subfamilies, of which the romance languages (French, Italian, Spanish, Catalan Romanian, Portugese, plus a few less well-known languages, such as Romansch,

Sardinian, Galician and Occitan) and Germanic (German, English, etc.) are two of the largest. Others such as Armenian and Albanian have only a single member. Latin, extinct but still widely studied and taught, belongs to the Romance family and indicates how such families have numerous extinct forebears. The largest subfamily of Indo-European is the Indo-Iranian family, which includes Sanskrit and the numerous languages of the Indian subcontinent and Iran.

But the branching of the language tree does not stop here. The Celtic language part of the tree is split into Welsh (still spoken in parts of Wales with a strong cultural tradition), Gaelic (still spoken in Ireland and Scotland and the Isle of Man), to virtually extinct languages such as Cornish and Breton. However, in even these languages communities are actively trying to invigorate and extend their use.

The classification, like that of the biological world, is *genealogical.* Languages are linked together by common heredity. It is also possible to classify languages according to types of structural features they exhibit, such as the linking of concepts in the word for firework display we saw earlier. The world's languages are divided into three broad categories: *inflectional*, where words change in fixed ways to define shades of meaning, as in the case of forming singular and plurals of nouns; *isolating,* such as Chinese, where each word is a morpheme (the smallest unit of meaning); and *agglutinative,* such as Turkish, where many morphemes may be joined together to form longer words [29]. This aspect of classification concerns us here in only one way—the need for *stemming,* i.e., taking out the grammatical components to leave just the basic unit such as *run* from *running, runs, ran.*

One might ask why it should be worth preserving these languages at all. English is now the most widely used language as either first or second language and dominates the Internet and much scientific discourse. Among the many arguments for documenting and preserving cultural artefacts, the case for languages is that it provides unique insight into cultures and ways of thinking. It is this element which is especially important to this book.

Agent modelling of social systems requires a detailed understanding of how agents behave. To gather this information, particularly for remote or isolated communities, it is essential to develop an accurate understanding of the language used to describe a variety of scenarios. In effect, one needs to build an *ontology* of the model target, be it water use in Bali water temples or fishing in island communities [21]. Such an ontology then becomes valuable at the modelling level in helping to define the rules which define agent behaviour. Section 3.2.2 takes up the influence of language on thought. Section 3.4 discusses ontologies in more detail.

3.2.2 *The Language Glass*

In English we often talk about the banks of a river in terms of their geographical location, similar to the aboriginal tribes above. So we talk about the South Bank of the River Thames and so on. In France, however, the approach is relative. Imagine facing down the river in the direction of flow, i.e., towards the mouth. Then one bank will be on the left, *la rive gauche*, and the other on the right, *la rive droite*. Since the Seine winds its way through Paris in a highly convoluted way, this nomenclature is much easier to handle unless you have an in-built compass!

In the early days of linguistics, Benjamin Lee Whorf [7] advanced the idea that we cannot have a concept without a word, that language was intricately tied up with how we perceive the world. The strong form of this hypothesis, that only some languages can express some ideas, is demolished by Guy Deutscher [20] in his book *The Language Glass*, but he nevertheless finds examples where language does in some ways condition thought, particularly where language makes transmitting some information mandatory. In Japanese, for example, the word *to give* has three forms depending upon the relative status of the recipient (inferior, equal or superior). Thus the choice of every word carries with it the assumption of the speaker (or at any rate the viewpoint the speaker wishes to convey) about the status relationship.

Two domains have proved particularly useful in studying the link between perception or cognition and language, because they admit of precise measurements. Words for colour (Section 3.2.3) vary considerably across languages, but colours for a given illumination can be represented precisely as wavelengths, or combinations of wavelengths of light. The other area which has attracted a lot of interest is counting and numerosity (Section 3.2.4).

3.2.3 *Perception of Colour*

Deutscher [19, 20] discusses how perception depends upon whether we have words for particular shade of colour—the blue-green boundary varies in different cultures. A recent elegant experimental result shows how perception can be influenced by language in the right half of the visual field. Gilbert et al. [32] showed that reaction times were faster in the right visual field when target and distractor colours in their experiment had different names. The right visual field feeds the left hemisphere of the brain, the hemisphere where language is normally predominant. But there was no difference in reaction time in the left visual field (corresponding to activity in the right hemisphere). Gilbert et al. [32] suggest that their results might arise from either or both of a direct effect of language on early visual processing or an effect at the higher-level decision-making stage.

The effect of language goes deep into early visual processing. Using EEG, Thierry et al. [64] showed that for Greek speakers who have different words for two shades of blue, not distinguished in English, there is a brain potential that appears very rapidly (within 100 ms) in the Greek speakers for stimuli differing in these shades. This very rapid onset is typical of processing in the very early stages of vision, in effect before language processing occurs.

An interesting refinement comes from similar sorts of experiments, but this time using invented words for new colour boundaries. After training, reaction times similarly improve across these boundaries. Thus the language influence on perception is learnable rather than innate [71].

Of course children are not born with their language skills intact. Thus one might ask if these colour effects appear before children have learned a language. Franklin et al. [30] found that in prelinguistic infants, the colour category perception is stronger in the *left* visual field, corresponding to the right hemisphere. They concluded that it is language that imposes the categories on the left hemisphere.

3.2.4 *Numbers and Words*

In some languages there are no specific words for numbers more than a just a few, such as no words to distinguish 8 from 9. In some cases there are words for one, two and many, or perhaps, including three [6]. Thus, various studies have looked at indigenous peoples where their language has had little contamination from languages of the developed world, where there are full sets of counting words.

Studies with Amazonian Indians [34] have implied that without words, comparing groups of different size or performing simple arithmetic operations is difficult, supporting the Whorf hypothesis [7]. Other work with Australian Aborigines [6], again with limited number words, does not find numerosity limitations and suggests that counting mechanisms precede language.

A possible reconciliation of these findings lies in another discovery of the difference between linear and logarithmic mappings. The Munducuru Amazonian populations used a logarithmic mapping which blurs the differences between larger numbers on the linear scale we are used to. In fact this tendency towards a logarithmic assessment of size might be the innate character with the linear scales learned formally [18]. Thus the area is somewhat controversial. The nature of the tasks and the setup of the experiments may be crucial. Further developments on this exciting topic are eagerly awaited.

3.3 Large-Scale Semantic Networks

The elements of languages are connected together in different ways. Words such as *red, crimson* and *scarlet* are closely related colours. Words such as *at, from,* and *with* are all prepositions and related grammatically. Thus we can build language networks using links between elements.

There are two broad approaches to building semantic networks [61]:

1. Use Thesaurus or Wordnet for links (Section 3.3.3).
2. Use collocation in a large body (corpus) of text (Section 3.3.5).

3.3.1 *Semantic Relatedness*

The nature of the links in language networks vary in several ways. In the syntax networks, grammatical relationships define links. In the semantic world there are several relationships:

Synonym a word with similar meaning to another word
Antonym a word with opposite meaning to another word
Hyponym a word which is a subclass of another word. *Jaguar* is a hyponym of *cat*
Hypernym a word which is a superclass of another word, essentially the opposite of hyponym. *Cat* is a hyponym of *jaguar*
Homonym a word which sounds the same as another. *Lead* is a metal, but *led* is the past participle of *lead*, but they sound the same
Polyseme a word having more than one meaning. *Jaguar* is a South American big cat and also an English make of car (Section 3.3.4)

Semantic relatedness does not have an exact definition, but it has become of ever-increasing importance, driven by the need for better and better search engines on the Web. One of the first drivers of the study of relatedness came from the idea of latent semantic indexing (Section 3.3.2) and how children acquire vocabulary.

Learning a lot of words is complicated by their frequency of occurrence. Not all words are equally likely, and one of the first statistical observations of language was Zipf's law [72]. Zipf found that the frequency of occurrence of words followed a power law. An interesting complement to Zipf's law is referred to as Heap's, or sometimes Herdan's, law. If words are collected by scanning documents, the number of novel words grows as a sublinear power of the number of words scanned [9]. Much more recently interesting statistical properties of word order have emerged. Montemurro et al. [47] found that across a range of languages from completely different families, there is universal entropy to word order. Languages vary in the numbers of words and how these words are ordered grammatically. But the relative entropy, the effective information in the word order alone is a constant across all these diverse languages.[a]

3.3.2 *Latent Semantic Indexing*

Latent semantic indexing arose from the realization by Landauer [42] that there was simply no way that children could grow their vocabulary at the observed rate by speech or learning vocabulary. Young adults reach 20,000 words in the course in less than 20 years, meaning an average of 1,000 words per year, or 20 words per week. They are unlikely to see that number of different words from spoken language, so this growth has to come from reading.

Landauer thus realized that the meanings of many words must be absorbed through context rather than specific definition. At the time computational resources were limited in what they could do, but since then there has been an enormous growth in analysing text for word semantics. This takes several courses:

1. Analysing text to infer relationships between words
2. Building concept maps or ontolgoies for specialised domains. Leximancer, discussed in Section 3.4.2, is one the primary tools for doing this
3. Inferring aspects of cognition. This can vary from establishing authorship and plagiarism to inferring dementia, as in Peter Garrard's seminal work on Iris Murdoch (Section 3.4.2)

3.3.3 *Thesauri and Wordnet*

Roget's Thesaurus has been with us for a long time and provides a comprehensive index of synonyms and antonyms. One can use the links between words found in the thesaurus to generate a semantic network. Motter [48] did precisely this. Over two orders of magnitude they find a reasonable approximation to a power law, as shown in Fig. 3.2.

---

[a] Technically the Kullback Leibler divergence between distributions of text sequences drawn from language text and randomized versions thereof.

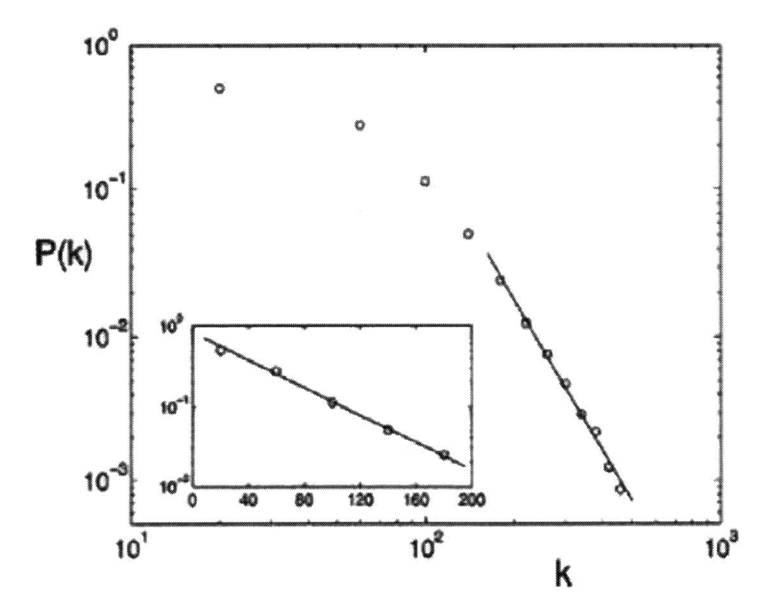

Figure 3.2 Language network from thesauri (from Motter, see text). The inset shows an initial exponential fall-off.

3.3.3.1 Wordnet analysis

Wordnet [25, 45] is a major initiative to build language trees and has been used for a variety of measures of semantic relatedness [5]. It consists of nine major synsets of words which bear some relationship with each other. It was constructed by linguists and contains a great deal of human analysis and judgement. The first Wordnet was in English, but there are now other variants, such as Germanet for German [8].

Wordnet is essentially a forest in which the nine synset categories represent trees. There are a few links between trees, meaning that the tree model is not completely accurate. Analysing link distributions for Wordnet produces a very clean scale-free distribution, with exponent around 3 [63].

$$s(c_1, c_2) = \max_{c \in S(c_1, c_2)} i(c) \tag{3.1}$$

where $S(c_2, c_2)$ is the set of concepts subsuming c_1 and c_2 and $i(c)$ is the information content of concept c. Resnik's definition was simply based on the probability of finding the concept within a corpus (Eq. 3.2).

$$i_{\text{res}} = -\log p(c) \tag{3.2}$$

Seco et al. [58] defined a new information measure (Eq. 3.3 based on Wordnet), arguing that this was easier to calculate and uses a standard source (although Wordnet may continue to evolve). Resnik [56] and others calculated similarities from corpora [37] or taxonomies such as Wordnet and compared the values so obtained with experimentally obtained human judgements of the similarities between words.

$$I = \frac{1 - \log(h(c) + 1)}{\log(\max_{\text{wn}}^{\text{wn}})} \tag{3.3}$$

Such estimates of distance between concepts do not really mean anything unless we can relate them back to human judgements. There have been several attempts to do this, but it is inevitably difficult to get a lot of judgements. So the comparison with human data is at best limited.

Resnik carried out partial replication of a canonical dataset produced much earlier by Rubenstein and Goodenough [57] which used 51 subjects to judge the similarity of 65 word pairs. Since then, other methods of generating word pairs have been developed and further sets of human judgements compiled [40, 69]. In fact entire workshops are now devoted to measures of linguistic distance. No one index comes top in all cases, but the Jiang and Conrath measure [37] is consistently near the top. Their distance, d_{jc}, on Wordnet is closely related to Resnick and Seco, using the occurrence probabilities of the concepts c_1, c_2, and their nearest common ancestor, c (Eq. 3.4).

$$d_{\text{jc}} = 2\log -(\log c_1 + \log c_2) \tag{3.4}$$

To have metrics among concepts is obviously very useful for search and text analysis algorithms, but it may not be theoretically sound. Concepts may not fit in a metric space [63]. In any metric space (such as the real world) distances have to have a property known as the *triangle inequality.* This means that for any three points in the space, a, b, c, the sum of the distances from a to b and b to c has to be greater than the distance from a to c as in an ordinary triangle. Concept distances may violate the triangle inequality. Polysemy makes this quite likely. Consider the set *chip, potato, computer. Chip*, with its dual meaning, is quite close to both *potato* and *computer*. But *computer* and *potato* are a long way apart.

The Seco et al. [58] approach was at least as good and sometimes slightly better than the other studies. The Seco distance depends strongly on how far up the Wordnet tree one must go to find a common ancestor. This is an intuitive match to a model of memory from Ross Quillian in the late 1960s, arguing from computational grounds. In a computer program, particularly within an object-oriented paradigm, information tends to be stored at an appropriate level of generality. Thus mammals have hair, cats are carnivorous mammals, tigers are cats, and cats have fur. To find out that a tiger has fur means going up two levels, but to find out that it is a carnivore means going up only one level. Subsequent psychological experiments found a difference in reaction time dependent on the number of levels which had to be transversed [14].

Since this formative work, there has been much interest in computing concept similarities and there are now commercial tools available, sometimes with proprietary metrics.

3.3.4 *Polysemy*

Polysemy is a particularly interesting feature of many languages, but it makes text analysis tricky. Co-occurrence of words can lead to unreliable estimates of similarity when the same lexical item can have more than one completely different meaning.

But polysemy may have definite value in human language. It has been argued that it can be a source of creativity, leading to unexpected links. But it might also be a key factor in efficient communication (Section 3.5).

3.3.5 *Semantic Networks*

The other approach to inferring language networks is collocation. Two words are linked if they appear together according to some metric. Despite numerous studies there is no unique way to do this. To begin with, there are many choices of text corpora. They can be fixed resources, such as the works of Shakespeare. They can be dynamic resources, of which one can take a snapshot, such as Wikipedia. But there have been attempts to use Google for estimating word similarity, where not only the resource (the Web) is changing, but it is effectively impossible to take a snapshot.

The unit of text in which the words must occur needs definition. It could be a window of a few words, a syntactic unit such as a sentence or paragraph or some other type of window. Given the co-occurrences which have been found, some sort of normalization is then required, and this, too, is not unique. We could divide by the total number of occurrences the product of occurrences of each and numerous other possibilities.

A semantic distance which matches human estimates of similarity would seem to be ideal, but there are difficulties here too. The examples given in Section 3.3.3.1 of necessity have very few word pairs and human judgements are themselves not unique.

Despite the variations in method, there seem to be some general results on the structure of human language networks.

3.3.5.1 Syntactic scaling

The concepts of a language, represented principally by nouns and verbs, are one sort of language dependency with which to create networks, in this case semantic networks. Links in the network arise when words have similar meaning. Ferrer et al. [27] examined a different network, one based on syntax.

The links now are based not on semantics but on syntactic relationships. These are directed, from a modifier word such as a noun to a head word such as a verb. Figure 3.3 shows how syntax networks are constructed from sentences following the Ferrer et al. paper. There are now two distributions of node degree, the in-degree distribution of links coming in and the out-degree distribution of links going out.

When language networks of this kind are constructed, several characteristics are shared with the semantic networks. For Czech, Romanian and Dutch the networks are all scale free with very similar exponents for both the in-degree and out-degree distributions. They all exhibit strong clustering and are disassortative. The scale

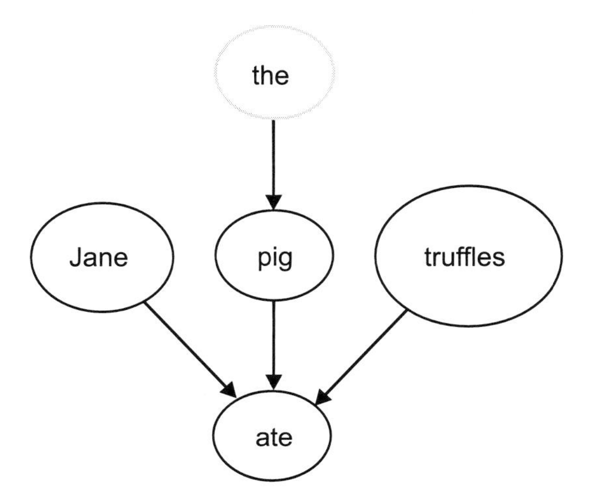

Figure 3.3 A syntactic network based on the sentences *Jane ate the pig* and *The pig ate truffles*. The graph is directed with modifiers pointing to head words (after Ferrer, see text). The article is shown here, but the connections are so great to particles of this kind that they would distort network statistics in general.

exponent is around 2 or a bit over and the between centrality follows an approximately inverse square law.

3.3.6 *Phonological Scaling*

Instead of written words, one can also look at phonological networks. These display some of the characteristics of the textual networks, but are not identical. Arbesan et al. [2] examined phonological networks across several widely different languages from English and Spanish to Hawaiian and Chinese. The nodes in a phonological network are still words, but the links show a strong similarity in sound (only a single phoneme different as in *hand, send, sad, sand*).

The previous networks have been practically fully connected, but the giant component (see Chapter 2) is much smaller, representing not much more than half the nodes. The scale-free character is weaker too, exhibiting an exponential cut-off as shown in Eq. 3.5.

$$P(z) = z^{-\alpha} e^{-\frac{z}{z_c}} \tag{3.5}$$

However, the networks do show strong assortative mixing, robustness and high clustering coefficients, these being higher/ greater than displayed in the semantic and syntactic networks.

3.4 Ontologies and Concept Mapping

Wordnet is a carefully constructed language taxonomy, with concepts arranged in suitable hierarchies, reflecting the usage of words in English (and now some other languages). For the World Wide Web and for building agent-based models, natural language understanding is still a grand challenge. Two techniques attempt to fill this gap in our understanding and machine technologies, which we consider in this section.

3.4.1 *Ontologies*

In natural language we often have many words for the same thing, or very slight variations on a concept. This huge diversity frustrates searching for information on the Web, which was one of the driving forces behind Web 2.0, the *semantic web*.

An ontology is generally a tiny subset of natural language describing a particular domain, say, *motor vehicles*. But the terms which describe an ontology themselves are diverse and duplicative. The core of an ontology are the classes, such as *cars, lorries*. They may have subclasses, such as *sports cars*. Classes bear a very close resemblance to classes in object-oriented programming but are also referred to as concepts. Classes have *attributes,* referred to also as *slots, roles,* or *properties,* such as *engine size*.

The World Wide Web Consortium developed a specification OWL, the Ontology Web Language [65], for precisely defining ontologies in a way which is machine readable. Not surprisingly the syntax structure is XML, relying on XML-based frameworks, such as the Resource Description Framework [66], for providing information about documents and other media resources. Machine-readable ontologies provide support for searching, organizing and mining documents.

Building an ontology is to a great extent a human task, requiring domain experts to carefully tease out relationships and find the best taxonomic structure. Should one classify a convertible as a sports car? If so, sports cars can have more than two seats and so on. But the task can be simplified, or at least begun, by using text analysis tools.

3.4.2 *Language Tools*

If we have a large body of text, then the techniques for determining language networks can be used to build concept maps, the precursors to ontologies. There are a variety of tools out there to do this, of which Leximancer [59] is a good example. Using methods similar to the language network studies described above, a map is drawn between words (concepts).

Figure 3.4 shows the result of applying Leximancer to a book on crisis communications [15]. The text is processed, common, everyday words, and grammatical entities such as prepositions are removed, leaving a map of concepts and the dominant themes that emerge.

Tools such as Leximancer are valuable for mapping the concepts (and words used to define them) in a body of text. This forms a cornerstone of qualitative research where understanding a culture, organization or social mechanism requires building such a concept map.

However, studying bodies of text has other uses. Authorship and plagiarism are sometimes detectable by looking at statistical properties. Another successful application was by Peter Garrard, who studied the works of the novelist Iris Murdoch, who, unfortunately, contracted Alzheimer's disease [31]. Signs of change in vocabulary use and language structure were evident in her last novel.

3.5 Language Evolution

When Chomsky [13] introduced the idea of universal grammar, a controversy began which still rages to some extent today, but progress has been made on several fronts. Section 3.1.5

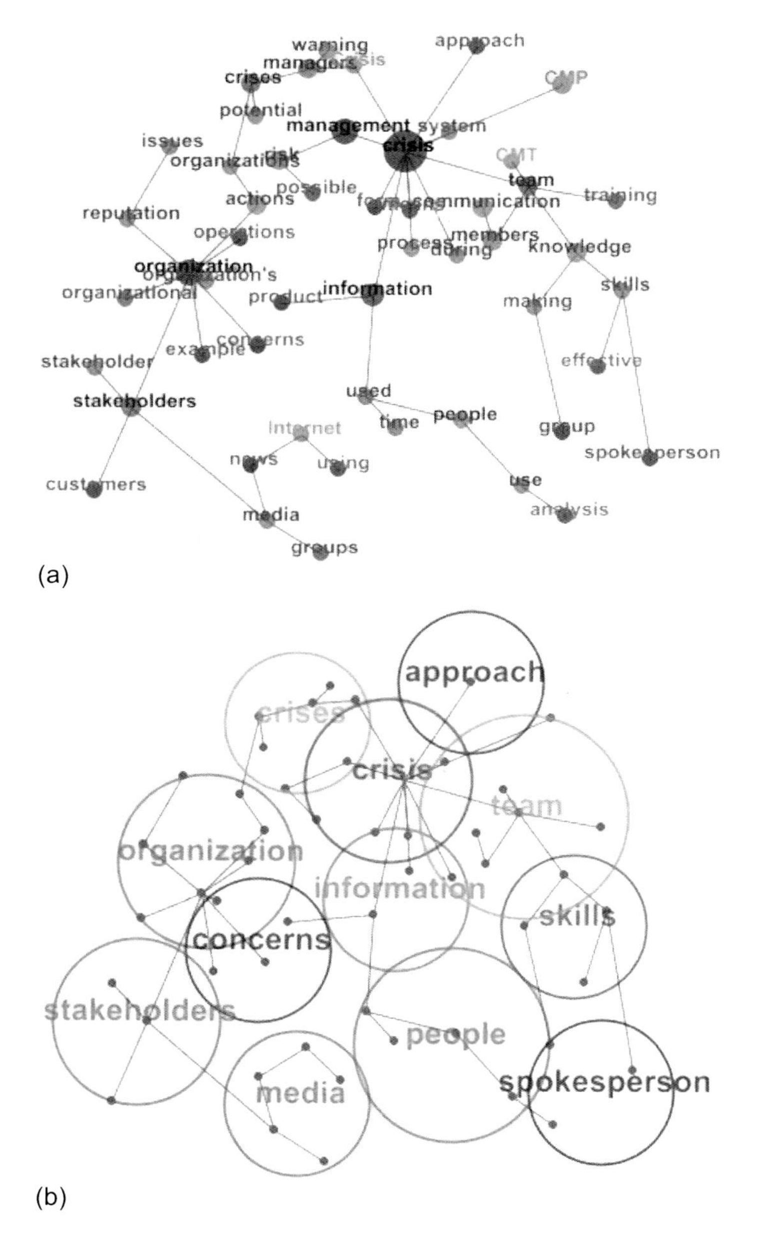

Figure 3.4 Concept map of crisis communications. (a) The concepts and the links between them. (b) The themes which emerge and visualize their interaction.

discusses some of the things which have been discovered about the implementation of language in the wetware of the brain. But there has also been progress on the theoretical front, discussed for example in the book by Martin Nowak [51].

It turns out, according to Nowak, that universal grammar is a theoretical necessity, because the class of human languages is not learnable from just hearing the language, without some assumptions about the rules of grammar. The form that this might take in the brain, however, is still very far from understood.

Thinking about the evolution of language as a communication system has proved to be very productive. Communication among mammals gets more sophisticated as we move towards primates and cetaceans such as dolphins. Monkeys have a wide variety of sounds, having different sounds to warn of a snake in the grass or an eagle overhead. Chimpanzees, grey parrots and dolphins seem to be able to learn something of human grammatical syntax, although the extent to which their own communication has any sort of sophisticated syntax is still an open question.

So, it is interesting to ask what drives the transition to grammar. As the number of possible situations increases, so does the advantage of having a grammar. Without grammar it is necessary to learn every possible case (*snake, eagle, leopard*...). Nowak [50] then provides a simple case estimate of the size of the language for syntax to be evolutionarily favoured (Eq. 3.6).

$$n \approx \frac{3}{p} \tag{3.6}$$

where n is the number of nouns and verbs and p is the probability that a particular combination of noun and verb occurs in the world. He also finds that the number of words, $n_{\max}$ that can be maintained in a language, using Zipf's law of word frequency, is

$$n_{\max}(\gamma + \ln n_{\max}) = bq \tag{3.7}$$

where γ is Euler's constant (0.5772...), q is the probability that a word will be learned on a single exposure and b the number of word learning events. Naturally, Eq. 3.7 contains some simplifying assumptions, but it does demonstrate that the number of words which can be sustained in the population depends upon the number of learning events (i.e., length of exposure, language learning time, etc.) to which people are exposed.

Finally an elegant, artificial life model by Ferrer i Cancho and Solé [26] gives a tidy demonstration of power law scaling in language. Their model assumes that communication is a balance in the mapping between a set of n signals, S, $\{s_1, s_2, \ldots s_n\}$ describing a set of m objects, R, with probability distributions on each.

The least effort for the sender is to have the smallest entropy, $H(S)$ in S while the receiver wants the lowest possible conditional entropy, $H(R|S)$ (which means that the chance of making an error is lowest). This leads to an energy function for the communication system given by Eq. 3.8

$$\Omega(\lambda) = \lambda H(R|S) - (1 - \lambda)H(S) \tag{3.8}$$

The outcome is a phase transition at a particular value of λ, where the *mutual information* (Eq. 3.9) between sender and receiver is maximal. At this point, characteristic of phase transitions in physics [67], power law behaviour appears and Zipf's law for word frequency falls out naturally.

$$I(R, S) = H(R) - H(R|S) \tag{3.9}$$

Artificial life models, studying the evolution of language and communication are proving particularly fruitful at the present time. Category games attempt to look at ways of assigning arbitrary names understood by most people within a population. The number of categories turns out to be small [53] and finite, which is a feature of human languages (such as the relatively small number of common terms for colour). The convergence to these categories has interesting dynamics, where the final smallish number of states is only weakly dependent on the size of the population of speakers [49]. Moreover, the structure of the categories reflects human sensitivity to different wavelengths of light [3].

3.5.1 *Language Competition*

Language evolution has some other parallels with the evolution of biological species [60]. Apart from change, life and death, languages exhibit competition among each other. In some areas of the world, more than one language can coexist. In Barcelona, for example, both Catalan and Spanish are spoken and many road signs, labels in shops and so on are bilingual.

Where multiple languages coexist, a useful idea has been that of language status: one language may be perceived as more important than the others, perhaps because it is the language of administration, or perhaps because it is the language of the aristocracy or the ruling class, as with French in 19th-century Russia. Abrams and Strogatz [1] developed a simple model for language competition using status as a key variable. Their result was not very encouraging: the higher status language would tend to wipe out the others. They suggested that only determined social policies, such as the maintenance of French, this time the Canadian underdog, in Montreal.

The preservation of languages, as we discussed above, has a degree of urgency about it, and hence quite a lot of effort has gone into refining the Abrams and Strogatz model. One generalization is to include a sub-population of bilingual speakers. Under some conditions this will not allow both languages to survive, providing they are sufficiently similar [46]. Another is to allow the prestige to vary, which also allows for both languages to survive under some parameter settings [11].

3.5.2 *Cortical Scaling Across Species*

Language has strong network characteristics, thus one is tempted to ask whether or not these networks are reflected to some extent in the structure of the brain itself. This is far from answerable at the present time, but there are a few interesting results to conclude the chapter.

One of the most interesting ideas relating brain, language and society appeared in the book *Gossip, Grooming and the Evolution of Language*, in which Robin Dunbar [22–24] argued that the leap in brain size seen in *homo sapiens* arose from the need to maintain relationships in larger groups and these were in turn mediated by language. Numerous other studies [10, 28, 68] describe these ideas.

The great deal of interest sparked in networks over the last decade has also led to a number of studies of networks in the brain, not just at the local level, but in the brain as a whole. Section 3.5.5 summarizes the state of the art. Some broad features of the brain *are* similar to the language networks we have looked at, but the details do not match up particularly well.

However, if there were to be any relationship between brain organization and language structure, then one would expect to see some consistent language patterns in the brain across multiple individuals. The areas of the brain associated with language have been known for a long time, but can be drilled down to more specific details such as concept or grammatical categories. Section 3.5.4 describes evidence which suggests that to a limited extent this *is* possible.

While neuroscientists are collecting data about the organization of language in the brain, linguists and computational intelligence researchers are interested in models for language learning in neural networks, the so-called connectionist approach. Section 3.1.5.4 describes how these approaches can capture some of the characteristics of how language degrades during dementia or injury. The understanding of the relationship of language and the brain is still at a very early stage. There will be many exciting results ahead.

3.5.3 *The Social Brain*

There is an approximately linear log–log relationship (i.e., a power law) between the size of the neocortex and group size across a range of primates [23]. The noecortex is the most recently evolved part of the cerebral cortex responsible for many higher functions, including language. Humans, of course, have the largest neocortex and *also* have the largest group size.

Group size across a wide range of human societies and structures is remarkably consistent reinforcing the conjecture that it is the brain size which determines the group size. Dunbar speculates that the need to maintain the relationships and interactions among group members needs a larger brain as group size increases. Since the number of possible interactions increases as the square of the group size, it is easy to see why this might be a power law growth.

Some primates may occupy similar ecological niches and be quite similar in size, diet and other characteristics. Yet some form coalitions within their groups and others do not. There is a striking correlation with neocortex size, the animals which form coalitions having the larger neocortex.

With the rapid rise in social media (Facebook, Twitter and the like), the question arises as to whether effective group size is actually increasing. This is an open research question, but some results already suggest that the core group for individuals has not increased significantly in size [33].

Conversely, one might ask if the much larger groups with which people can associate will drive brain evolution or at least select for traits which help in the management of large groups, what one might call tongue-in-cheek the political genes. Conversely, it may be that systems such as Facebook which are computer network mediated may ultimately be able to store information outside the human brain. It is certainly possible to keep track of what one has said to whom through computer-mediated messaging. But the computational memory architectures, both hardware and software, available now do not have the flexibility to carry out the numerous links, cross-validations and insights of which the prefrontal cortex is capable.

3.5.4 *Categories in the Brain*

Current medical imaging techniques are a long way from resolving down to the individual neuron level. Single neuron results are almost always obtained from electrophysiological studies on animals. So we are still a long way from linking language categories and networks to brain architecture, if indeed, there is any fine-grained link. But some tantalizing results are already in.

Hasson et al. [36] carried out fMRI imaging[a] on subjects watching a Clint Eastwood movie [36]. Frames were selected for their focus on buildings, faces and hands, and neuroimaging revealed distinct brain areas for each of these categories.

Thus there is broad separation of different objects into different brain areas. How far this can be pushed, into finer and finer

[a]fMRI measures the spatial distribution of blood flow within the brain. It has high spatial resolution, but low temporal resolution, since increased blood flow lags behind activity.

subcategories, is an open question. Here an old idea has gained some new credibility. In the early days of brain science people talked about a grandmother cell, a cell which fired only in response to the grandmother. But this was soon discredited, with consensus that representations were distributed over many neurons.

Recently this connectionist viewpoint has come under renewed scrutiny. Quiroga et al. [54] found examples of neurons which were *very* selective for celebrity figures such as Bill Clinton or Halle Berry. These cells were sensitive not just to an image, but to all sorts of other indicators, such as voice, with a wide variety of possible stimuli across each modality. In other words, they seem to respond very closely to the concept however it might be activated. A Bill Clinton cell would be very inactive in response to a picture of Halle Berry, and vice versa.

These results are about concepts rather than about words, since they span multiple modalities, although the two are intricately related. But they raise the question of what happens if somebody is multilingual. Different parts of the brain might handle different languages, and this would seem to be quite likely for somebody who grows up monolingual and acquires other languages at school or in adulthood. But people who grow up multilingual will learn words for, say, cat in different languages at the same time. Evidence suggests that the concept of cat exists in one place with subdivisions for different languages. This seems to fit in well with the Halle Berry neuron results.

Perani and Abutalebi [52] further differentiate a second language (L2) learned later in life with early bilingualism. It seems that as proficiency increases, the areas of brain activity merge. But where the second language still lags behind the first, additional brain areas are used in processing L2 compared with L1.

Crinion et al. [16] looked at semantic priming, an experimental psychology procedure wherein a priming word will reduce the reaction time to a related word, such as the combination trout-salmon. They showed that if the priming word is in a different language, e.g., *Forelle*, the German word for trout, then cortical activation remains very similar and for bilingual Germam-English speakers there is no reaction time difference. The overarching conclusion was that the brain areas for the two languages are intermingled and concept driven. However, one particular brain area, the *left caudate nucleus*, was more active when the language of the words in the word pair was different. Thus this seems to be a control area, which manages selection of words from the correct language.

Current opinion now seems to favour both L1 and L2 being active at the same time with dynamic inhibition. In fact, in a recent radio discussion, Judith Kroll reports that in some Spanish/English bilinguals, where English has been learned later in life, L1 shifts to English in neurological signature [41]. From an evolutionary perspective, Hagen [35] argues that learning the first language is highly advantageous, and thus it is not surprising that it happens fast and effortlessly early in life, but that the advantage of a second language is much weaker. Thus it is not surprising that it requires much greater effort to learn a second language later in life. A naturally corollary would be to find different brain structures involved in second-language processing.

3.5.5 *Characteristics of Brain Networks*

Sporns et al. [62] surveyed a range of studies of networks in the brain from anatomical studies on rat and cat to fMRI studies of the human brain [12, 43, 70]. Two common patterns emerge: a scale-free architecture; and high clustering coefficients. Both these characteristics would be what we would expect at a very general level in the cortical implementation of language networks.

Some neurons in the brain have a *myelin* sheath around them. It is a white fatty substance, which gives rise to the term white matter. The grey matter neurons are unmyelinated. Impulses travel a lot faster over myelinated neurons, and thus the white matter forms the long-range connections of the brain. As the brain gets bigger, more and more of these long-range connections are required. Laughlin and Sejnowski [44] give a very tight power law relationship (Eq. 3.10), with the white matter scaling as the 1.23 power of the gray matter, $W \propto G^{1.23}$, where W and G are the white and grey matter volumes in mm^3, respectively. This relationship holds over five orders of magnitude from pigmy shrew at about 10 mm^3 to elephant at over a liter:

$$\log_{10} W = 1.23 \log_{10} G - 1.47 \qquad (3.10)$$

3.6 ENVOI

English is now dominant on the World Wide Web and is the most widely used of all languages if first and second languages are grouped together. Fischer [29] suggests that we may see a collapse to just a few, maybe even one language across the world, and that extinct languages can never really be revived.

With the loss of a language go all the specific cultural references and implications within it. Cultures merge, values become shared and the need for special vocabulary dies. Yet, at the same time, as in any ecosystem, new groups emerge with their own special linguistic variations. The richness of language and its ongoing variation and diversity underlies the identification and simulation of social networks.

References

1. Abrams, D. M., and Strogatz, S. H. (2003). Modeling the dynamics of language death. *Nature* **424**, p. 900.
2. Arbesan, S., Strogatz, S. H., and Vitevitch, M. S. (2010). The structure of phonological networks across multiple languages. *Int. J. Bifurcation Chaos* **20**, pp. 679–685.
3. Baronchelli, A., Gong, T., Puglisi, A., and Loreto, V. (2010). Modelling the emergence of universality in colour naming patterns. *Proc. Natl. Acad. Sci. USA* **107**, pp. 2403–2407.
4. de Bot, K., and Stoessel, S. (2002). Introduction: language and social networks. *Int. J. Soc. Lang.* **153**, pp. 1–7.
5. Budanitsky, A., and Hirst, G. (2006). Evaluating WordNet-based measures of lexical semantic relatedness. *Comput. Ling.* **32**(1), pp. 1–35.
6. Butterworth, B., Reeve, R., Reynolds, F., and D. Lloyd, D. (2008). Numerical thought with and without words: evidence from Indigenous Australian children. *Proc. Natl. Acad. Sci. USA* **105**, pp. 13179–13184.

7. Carroll, J. B., ed. (1956). *Language, Thought and Reality: Selected Writings of Benjamin Lee Whorf* (MIT Press, Cambridge, MA).
8. Carstenen, K. U., Ebert, C., Endriss, C., Jekat, S., Kabunde, R., and Langer, H., eds. (2004). Lexiaklisch semantische Wortnetze, in *Computerlinguistik und Sprachtechnologie* (Spektrum Akademsicher Verlag, Heidelberg), pp. 423–431.
9. Cattuto, C., Barrat, A., Baldassarri, A., and Schehr, G. (2009). Collective dynamics of social annotation. *Proc. Natl. Acad. Sci. USA* **106**(26), pp. 10511–10515.
10. Changizi, M. A. (2001). Principles underlying mammalian neocortical scaling. *Biol. Cybern.* **84**(3), pp. 207–215.
11. Chapel, L., Castello, X., Bernard, C., Deffuant, G., Eguiluz, V. M., Martin, S., and Miguel, M. S. (2010). Viability and resilience of languages in competition. *PloS One* **5**, e8681.
12. Cherniak, C., Mokhtarzada, Z., Rodriguez-Esteben R., and Changizi, K. (2004). Global optimisation of cerebral cortex layout. *Proc. Natl. Acad. Sci. USA* **10**, pp. 1081–1086.
13. Chomsky, N. (1972). *Language and Mind* (Harcourt Brace Javanowich, New York).
14. Collins, A. M., and Quillian, M. R. (1969). Retrieval time from semantic memory. *J. Verb. Learn. Verb. Behav.* **8**, pp. 240–247.
15. Coombs, W. T. (2007). *Ongoing Crisis Communication* (Sage Publications, Thousand Oaks).
16. Crinion, J., Turner, R., Grogan, A., Hanakawa, T., Noppeny, U., Devlin, J. T., Aso, T., Urayama, S., Fukuyama, H., Stockton, K., Usui, K., Green, D. W., and Price, C. J. (2006). Language control in the bilingual brain. *Science* **312**, pp. 1537–1540.
17. Crystal, D. (2000). *Language Death* (Cambridge University Press, Cambridge).
18. Dehane, S., Izard, V., Spelke, E., and Pica, P. (2008). Log or linear? Distinct intuitions of the number scale in Western and Amazonian indigene cultures. *Science* **320**, pp. 1217–1220.
19. Deutscher, G. (2006). *The Unfolding of Language* (Arrow Books, London).
20. Deutscher, G. (2010). *Through the Language Glass* (Heineman, London).
21. Dray, A., Perez, P., Jones, N., Le Page, C., D'Aquino, P. White, I., and Auatabu, T. (2006). The AtollGame experience: from knowledge engineering to a computer-assisted role playing game. *JASSS* **9**(1), pp. 1–11.

22. Dunbar, R. (1996). *Gossip, Grooming and the Evolution of Language* (Faber and Faber, London).
23. Dunbar, R. I. M. (1993). Co-evolution of neocortex size, group size and language in humans. *Behav. Brain Sci.* **16**(4), pp. 681–735.
24. Dunbar, R. I. M., and Shultz, S. (2007). Evolution in the social brain. *Science* **317**(5843), p. 1344.
25. Fellbaum, C., ed. (1998). *Wordnet: An Electronic Lexical Database* (MIT Press, Cambridge, MA).
26. Ferrer i Cancho, R., and Solé, R. V. (2003). Least effort and the origins of scaling in human language. *Proc. Natl. Acad. Sci. USA* **100**, pp. 788–791.
27. Ferrer i Cancho, R., Solé, R. V., and R. Köhler. (2004). Patterns in syntactic dependency networks. *Phys. Rev. E* **69**, p. 051915.
28. Finlay, B. L., Darlington, R. B., and Nicastro, N. (2001). Developmental structure in brain evolution. *Behav. Brain Sci.* **24**(02), pp. 298–304.
29. Fischer, S. R. (1999). *A History of Language* (Reaktion Books, London).
30. Franklin, A., Drivonikou, G. V., Bevis, L., Davie, I. R. L., Kay, P., and Regier, T. (2008). Categorical perception of color is lateralized to the right hemisphere in infants, but to the left hemisphere in adults. *Proc. Natl. Acad. Sci. USA* **105**, pp. 3221–3225.
31. Garrard, P., Maloney, L. M., Hodges, J. R., and Patterson, K. (2005). The effects of very early Alzheimer's disease on the characteristics of writing by a renowned author. *Brain* **128**, pp. 250–260.
32. Gilbert, A. L., Regier, T., Kay, P., and Ivry, R. B. (2006). Whorf hypothesis is supported in the right visual field but not the left. *Proc. Natl. Acad. Sci. USA* **103**, pp. 489–494.
33. Goncalves, B., Perra, N., and Vespignani, A. (2011). Validation of Dunbar's number in Twitter conversations. *PLoS One* **6**, e22656.
34. Gordon, P. (2004). Numerical cognition without words: evidence from Amazonia. *Science* **306**, pp. 496–499.
35. Hagen, K. (2008). The bilingual brain: human evolution and second language acquisition. *Evol. Psychol.* **6**, pp. 43–63.
36. Hasson, U. (2004). Intersubject synchronization of cortical activity during natural vision. *Science* **303**, pp. 1634–1640.
37. Jiang, J. J., and Conrath, D. W. (1997). Semantic similarity based on corpus statistics and lexical taxonomy. In *Proceedings of the International Conference on Research in Computational Linguistics*, pp. 19–33.

38. Ke, J. (2007). Complex networks and human language. http://arxiv.org/abs/cs/0701135.
39. Kenneally, C. (2007). *The First Word* (Penguin Books, New York) .
40. Klebanov, B. B. (2006). Measuring semantic relatedness using people and WordNet. In *Proceedings of the Human Language Technology Conference*, pp. 13–16.
41. Kroll, J. (2011). Your fabulous bilingual brain. http://www.abc.net.au/rn/allinthemind/stories/2011/3164263.htm (ABC Radio, Sydney).
42. Landauer, T. K., Foltz, P. W., and Laham, D. (1998). Introduction to latent semantic analysis. *Discourse Process* **25**, pp. 259–284.
43. Laughlin, S. B. de Ruyter van Steveninck, R. R., and Anderson, J. C. (1998). The metabolic cost of neural computation. *Nature Neurosci.* **1**(1), pp. 36–41.
44. Laughlin, S. B., and Sejnowski, T. J. (2003). Communication in neural networks. *Science* **301**(5641), pp. 1870–1874.
45. Miller, G. A., Beckwith, R., Fellbaum, C. D. Gross, D., and Miller, K. (1990). Wordnet: an online lexical database. *Int. J. Lexicogr.* **3**, pp. 235–244.
46. Mira, J., Seoane, L. F., and Nieto, J. J. (2011). The importance of interlinguistic similarity and stable bilingualism when two languages compete. *New J. Phys.* **13**, p. 033007.
47. Montemurro, M. A., and Zanetta, D. J. (2011). Universal entropy of word ordering across linguistic families. *PLoS One* **6**(5), e19875.
48. Motter, A. E., de Moura, A. P. S., Lai, Y.-C., and Dasgupta, P. (2002). Topology of the conceptual network of language. *Phys. Rev. E* **65**, p. 065102.
49. Mukherjee, A., Tria, F., Baronchelli, A., Puglisi, A., and Loreto, V. (2011). Aging in language dynamics. *PLoS One* **6**, e16677.
50. Nowak, M. A. (2000). Evolutionary biology of language. *Proc. R. Soc. Lond. NB* **355**, pp. 1615–1622.
51. Nowak, M. A. (2006). *Evolutionary Dynamics* (Belknap Press of Harvard University Press, Cambridge, MA).
52. Perani, D., and Abutalebi, J. (2005). The neural basis of first and second language processing. *Curr. Opin. Neurobiol.* **15**, pp. 202–206.
53. Puglisi, A., Baronchelli, A., and Loreto, V. (2008). Cultural route to the emergence of linguistic categories. *Proc. Natl. Acad. Sci. USA* **105**, pp. 7936–7940.

54. Quiroga, R. Q., Reddy, L., Kreiman, G., Koch, C., and Fried, L. (2005). Invariant visual representation by single neurons in the human brain. *Nature* **435**, pp. 1102–1107.
55. Oxford Media Release. Race against time and elements to record language. http://www.ox.ac.uk/media/news_stories/2011/112104.html, 2011.
56. Resnik, P. (1995). Using information content to evaluate semantic similarity in a taxonomy. In *Proceedings of the 14th International Joint Conference on AI*, pp. 448–453.
57. Rubenstein, H., and Goodenough, J. B. (1965). Contextual correlates of synonymy. *Comm. ACM* **8**, pp. 627–633.
58. Seco, N., Veale, T., and Hayes, J. (2004). An intrinsic information content metric for semantic similarity in Wordnet. In *Proceedings of ECAI*, pp. 1089–1090.
59. Smith, A. (2011). Leximancer. https://www.leximancer.com.
60. Solé, R. V., Corominas-Murtra, B., and Fortuny, J. (2010). Diversity, competition, extinction: the ecophysics of language change. *J. Roy. Soc. Interface* **7**(53), pp. 1647–1664.
61. Solé, R. V., Murtra, B. C., Valverde, S., and Steels, L. (2005). Language networks: their structure, function and evolution. http://www.santafe.edu/research/publicat...papers/05-12-042.pdf.
62. Sporns, O., Chialvo, D. R., Kaiser, M., and Hilgetag, C. C. (2004). Organization, development and function of complex brain networks. *Trends Cognit. Sci.* **8**, pp. 418–425.
63. Steyvers, M., and Tenenbaum, J. (2005). The large-scale structure of semantic networks. statistical analysis and a model of semantic growth. *Cognit. Sci.* **29**, pp. 41–78.
64. Thierry, G., Athanasopoulos, P., Wiggett, A., Dering, B., and Kuipers, J.-R. (2009). Unconscious effects of language-specific terminology on preattentive colour vision. *Proc. Natl. Acad. Sci. USA* **106**, pp. 4567–4570.
65. W3C. (2004). OWL web ontology language. http://www.w3.org/TR/owl-ref/.
66. W3C. (2004). Resource description framework. http://www.w3.org/TR/2004/REC-rdf-schema-20040210/.
67. Wicks, R., Chapman, S., and Dendy, R. (2007). Mutual information as a tool for identifying phase transitions in dynamical complex systems with limited data. *Phys. Rev. E* **75**(5), p. 051125.

68. Yopak, K. E., Lisney, T. J., Darlington, R. B., Collin, S. P., Montgomery, J. C., and Finlay, B. L. (2010). A conserved pattern of brain scaling from sharks to primates. *Proc. Natl. Acad. Sci. USA* **107**(29), p. 12946.
69. Zesch, T., and Gurevych, I. (2006). Automatically creating datasets for measures of semantic relatedness. In *LD '06 Proceedings of the Workshop on Linguistic Distances,* Association for Computational Linguistics, pp. 16–24.
70. Zhang, K., and Sejnowski, T. J. (2000). A universal scaling law between gray matter and white matter of cerebral cortex. *Proc. Natl. Acad. Sci. USA* **97**, pp. 105621–105626.
71. Zhou, K., Mo, L., Kay, P., Kwok, V. P. Y., Ip, T. M. M., and Tan, L. H. (2010). Newly trained lexical categories produce lateralised categorical perception of colour. *Proc. Natl. Acad. Sci. USA* **107**, pp. 9974–9978.
72. Zipf, G. K. (1965). *Human Behavior and the Principle of Least Effort: An Introduction to Human Ecology* (Hafner, New York).

Chapter 4

Complexity and Human Society

David G. Green and Suzanne Sadedin
Centre for Intelligent Systems Research, Monash University, Wellington Road, Clayton, Victoria 3800, Australia
david.green@monash.edu

Human society involves many networks of human relationships, as well as the interplay between people, institutions and processes. Social organization has evolved many ways of reducing and containing complexity, especially encapsulation, but several processes still lead to unplanned social trends. These processes include cascading contexts, positive feedback, changes in network connectivity, and brittleness of "divide and rule" strategies. Examples of these are evident in the social side effects of new technologies.

4.1 Introduction

Complexity is richness in the behaviour and organization of a system and emerges from interactions among its components. Social complexity arises from the ways in which people, institutions and entire societies interact with each other and with their environment.

Networks in Society: Links and Language
Edited by Robert Stocker and Terry Bossomaier

ISBN 978-981-4316-28-6 (Hardcover), 978-981-4364-82-9 (eBook)
www.panstanford.com

Social complexity often manifests itself in unexpected and sometimes sudden changes, such as outbursts of rioting, or rapid social trends. One of the prime concerns in studying social complexity is therefore how large-scale social processes and patterns emerge out of interactions between agents (individuals or organizations).

The need to understand social complexity is increasingly important in the modern world. Advances in technology, especially information and communications, are leading inevitably to increasing interconnectedness within modern society. Individuals can now communicate instantly with others anywhere in world. The words of a single person can influence people's thinking around the world. And actions in one place can set off chains of consequences that influence events in remote parts of the world.

Modelling approaches to social complexity have adopted a number of different paradigms over time, such as game theory and control theory. In more recent times, advances in computer technology have made it possible to simulate aspects of social systems directly. As a result, agent-based models (ABMs) are becoming increasingly used. ABMs simulate societies by representing each agent (person or organization) explicitly, as well as the processes involved [34].

In this account, we outline an interpretation based on complexity theory and, in particular, the network theory of complexity [24, 25, 27].

4.2 Emergent Social Structure and Behaviour

Some important aspects of social organization emerge out of direct person-person interactions. Here we look briefly at the kinds of collective behaviour and social networks organization. We draw an important distinction between these two phenomena. Collective behaviour concerns spontaneous outcomes that emerge from interactions between people. Social networks, on the other hand, concern social connections (e.g., family, neighbours) that lead to interactions.

4.2.1 *Collective Behaviour*

An important feature of complex systems is that system behaviour often depends not on the complexities of the individuals, but on interactions between the individuals. We see this most clearly in collective behaviour.

Sociological studies of collective behaviour have mostly centred around three general questions [43]:

1. How do people come to bypass/subvert institutional patterns and structures?
2. How do people come to translate their attitudes into significant overt action?
3. How do people come to act collectively rather than singly?

To address these questions, researchers have proposed a number of ideas about the processes involved. Contagion theory treats the spread of attitudes and behaviours as an epidemic process [9]. Game theory considers transitory, spontaneous behaviour in terms of the payoffs for individuals [8]. Threshold theory [23] holds individuals join an emerging movement only when a number of others are already engaged. This process, which is seen in the spread of new products, implies that people join a movement in some order, which depends on factors such as their dominance or independence.

The theory of emergent norms in collective behaviour suggests that peer influence leads individuals to align their opinions or behaviours with one another, resulting in the prevalence of a small number of norms of behaviour [57]. A commonplace example is seen when an audience spontaneously begins to clap in unison. We will examine the alignment of opinions in detail in Section 4.3.2.

The flashpoint model of social disorder [60] holds that a social group can reach a critical state, at which point any minor incident can suffice to trigger an explosive change in behaviour across the entire group. The Paris riots of 2005 provide an example. Two Muslim boys died accidentally while being chased by police. This incident sparked off riots among migrant communities. The rioting soon spread all over the city, then all over France, and even in to neighbouring countries. The violence continued for nearly a month [61].

Another example is well known from classical warfare. Military training conditions soldiers to respond as a group and to see safety in the group. This attitude is crucial in the face of imminent death on the battlefield. Direct confrontation between armies often led to a point where the confidence of soldiers in their fellows would falter. As panic set in, they would cease to think of themselves as a group and act as individuals. The influence of a few panicked individuals would shatter the waning confidence of others and a rout would ensue.

Groups emerge spontaneously when individuals align their behaviour with others around them. Simulation studies first demonstrated how this happens in insect societies ("stigmergy"). For instance, Hogeweg and Hesper [33] showed that bumble bees create complex social colonies when individual bees, following simple rules of behaviour, interact with each other, and with their environment.

It is easy to see stigmergy at work in the way self-organization occurs within an ant colony (Fig. 4.1). As ants wander around in their environment, they pick up any object they happen to come across and drop it again when they encounter a similar object. This procedure creates clusters of similar objects.

As clusters form, positive feedback comes into play. The ants continually shuffle material back and forth between piles. Random additions and removals causes larger clusters to grow at the expense

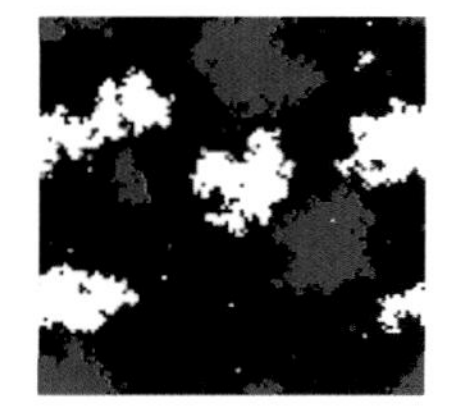

Figure 4.1 Formation of an ant colony by stigmergy. Objects in the area are shown as coloured spots. Ants wander around the area and pick up objects they come across. They drop an object they are carrying when they encounter another similar object. These actions, combined with positive feedback turns initially random scatter (left) first into clumps (centre) and finally into sorted piles of objects (right).

of smaller ones. The largest piles eventually absorb all the material from other piles, and in this way material within the colony is sorted.

Subsequent studies have shown that the same ideas apply in group behaviour. A flock of birds emerges when birds follow simple rules that coordinate the behaviour with other around them [51]. The same applies to schools of fish and herds of animals.

Models of human crowds have ranged between individual rule based models, such as those for animal groups, and particle models for very large groups. For instance, anyone who has experienced the movement of rush hour crowds at a busy railway station can appreciate the analogy of people with flowing streams of water. Simulations of crowd movement have often treated crowd movement of this kind as granular flows, especially in investigations of traffic and crowd motion. Simulations are used increasingly to identify large scale features (e.g., blockages) that can emerge in large groups of moving people. This is an important issue in building design, where the primary concerns are efficient flow and safety, especially around building exits (e.g., [40]).

4.2.2 *Social Networks*

Social networks are groups of agents (people, organizations) that are linked by social relationships or interactions. Unlike collective behaviour in crowds and other groups, which we looked at in the previous section, the links in social networks are not defined by the interactions, but are the channels by which interactions can occur. An obvious example is a family tree: individuals are the nodes and family connections (mother, daughter, etc.) define the links (network edges). Other common kinds of relationships include colleagues, friends and neighbours.

Networks of relationships between groups of people tend to form well-defined patterns. These topologies are discussed in detail in Chapter 2. They emerge as consequences of the way the network forms and evolves over time.

- ***Trees*** occur in two common situations. The first is via family descent: parent and child. Another is seen in management hierarchies, where one person supervizes another. Hierarchies are common in social organizations as a way to "divide and rule" large problems into smaller, more specialized roles.

- ***Small-world properties*** occur when individuals make "long-range" connections. For example, long-lived people or international travellers make connections between people otherwise widely separated in time or space.
- ***Scale-free networks*** occur when a network grows and edges form by preferential attachment. It is common in online social networks where people seek to make links with well-connected, popular individuals.

4.2.3 *Network Evolution*

Several kinds of processes govern the formation and self-organization of social networks. Several of these processes are well known in relation to the common network topologies discussed in the previous section. For instance, tree structures emerge in parent-child situations and scale-free structure emerges if new nodes attach preferentially to nodes of high degree. However, other processes also shape and modify social networks. External constraints are of course common, but are not complex in the sense we are considering. Here our prime concern is with mechanisms that emerge as a consequence of the complex network of connections. A good example is dual phase evolution.

Dual phase evolution (DPE) is a process by which repeated changes in the connectivity of networks (Fig. 4.2) allow variation and selection progressively to shape a network and its properties. Changes in the connectivity of social networks are common. In the course of their normal activity most people interact with relatively small networks of social contacts. In an office situation, for instance, people interact chiefly with their immediately colleagues. Likewise people mostly socialize with close friends. Sometimes, however, people attend conventions or go to parties. In doing so, they often make new social connections or become exposed to different ideas.

Dual phase evolution (DPE) has a different effect on a network from selection and variation acting at the same time. This is because of system memory: changes that emerge in one phase carry over

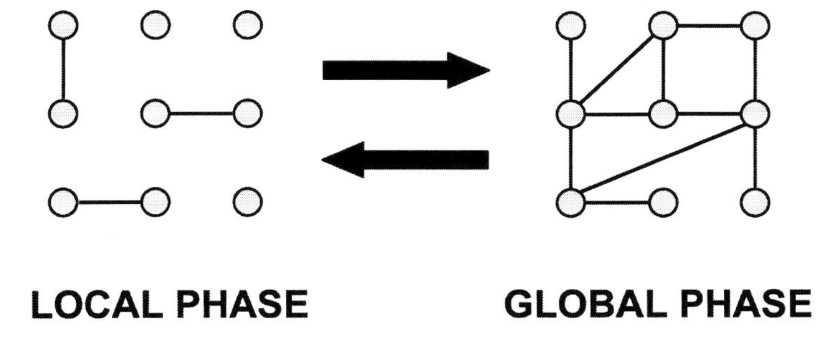

Figure 4.2 Dual phase evolution in a network. The system shifts repeatedly back and forth between the local and global phases. See text for further explanation.

into the other, so rather than cancelling out, selection and variation reinforce each other.

We can summarize the main points of the DPE model as follows:

1. The system has two phases. These arise from different edge densities in the system's underlying network.
2. Different processes act on the system in each phase. These occur in response to the changes in connectivity. One process enhances variation; the other reduces it.
3. The system changes repeatedly from one phase to the other. The changes may be stimulated by an outside influence.
4. The system has memory. The result of one process followed by the other is not the same as both processes operating simultaneously.

Models of two scenarios can serve to illustrate the role of DPE in two different social phenomena: the influence of media on public opinion [56], and the structure of social networks [44].

Field studies have shown that media can influence social behaviour [62].

The model of media influence (Fig. 4.3) assumes that people's activity falls daily into two phases: a social phase (selection) phase during the day when they are meeting with friends or colleagues and an isolated (variation) phase when they are at home being exposed to the media (e.g., watching TV, listening to the radio or reading the newspaper). Suppose that initially the population all

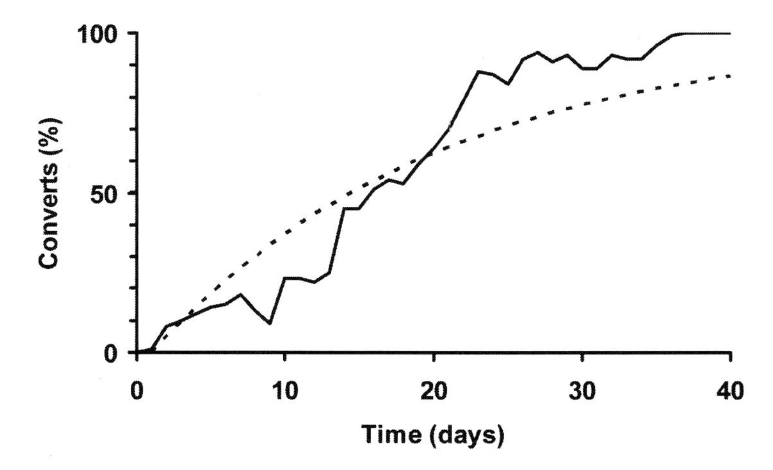

Figure 4.3 Simulation of the influence of media on public opinion. The graph shows the changes over time in the percentage of the population converted over time under 2 scenarios: (1) influence of the media alone (dotted line); and (2) dual phase evolution with alternating influence each day by media and peer (solid line). For further details see [57] and http://vlab.infotech.monash.edu.au/.

agrees on some issue (e.g., which brand of soap to use) but the media is promoting a different view. In the absence of peer influence, the proportion of converts in the population rises steadily until almost everyone is eventually converted. Under peer influence, however, the rate of conversion is initially much slower because the social network resists change. But when the number of converts reaches a critical level, the conversion rate increases rapidly. Ultimately the entire population is converted much faster than in the absence of peer influence. In the simulation scenario shown here (Fig. 4.3), converting the entire population required 119 days without peer influence, but just 37 days with it.

Dual phases also play a role in determining patterns of relationships within social networks (Fig. 4.4). Based on the idea of people in business spending most of their time in the office, but also going to receptions or conferences, this model assumes that people have two phases in their social interactions. Although social computing is changing the way people interact, this assumption is still valid for many parts of society. In a "local" or "selection" phase

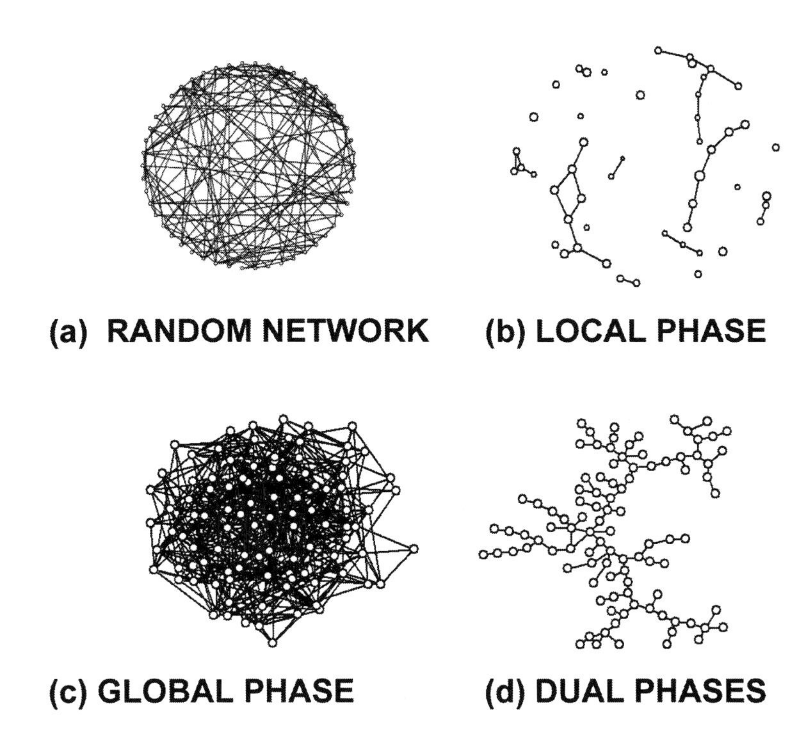

Figure 4.4 Formation of social networks by dual phase evolution (DPE). (a) an initial network with randomly assigned edges; (b) Result of evolution in local phase only; (c) Result of evolution in global phase only; (d) Result of DPE involving flips between local and global phases. Link formation during the global phase and link removal during the local phase are based on the numbers of characters that nodes share in common. See text for further explanation.

people interact only with their immediate contacts or neighbours whereas during a "global" or "variation" phase they interact with individuals from anywhere. The model further assumes that people make new social links only during the global phase, and break existing links only during the local phase (e.g., by lack of interaction). In the scenarios shown here (Fig. 4.4) individuals make links with others who are sufficiently similar and break links with individuals who are not similar enough. Models of this kind have been able to demonstrate the formation of cliques as well as the kinds of topologies (cf. Fig. 4.4d) found in studies of real-world networks [44].

4.2.4 *Consensus and Cooperation in Social Networks*

Social opinions, beliefs and norms of behaviour are often strongly influenced by interactions between individuals within a social network. As we saw above in the example of media and public opinion, peer influence plays a major role in setting social norms. An important question is how social interactions contribute to the emergence of consensus, cooperation and social order.

Many economists, sociologists and political scientists argue that trust and cooperation is rooted in self-interest [30]. Studies have shown that trust can have significant survival value [49] and experiments have shown that trust and cooperation are positively correlated [14]. Historically, societies have employed many practices to ensure trust and cooperation between individuals and groups. Examples include the use of hostages during political negotiations and the payment of deposits as a guarantee of intent to purchase.

One approach to the question of trust and cooperation has been to treat it from the perspective of game theory [1]. This interprets social issues in terms of strategies that individuals employ in their dealings with other individuals. The aim is to identify strategies (such as cooperation or defection) that become common and persist within a society.

Perhaps the best known example of a game theory approach is the Prisoner's Dilemma [6, 7]. Two prisoners have a choice: if they cooperate and do not inform on each other, then they will suffer a short jail term. On the other hand, if one of them defects and informs on the other prisoner, then the traitor gets set free, but his companion gets a very harsh sentence. The "game" is really an allegory for cooperation in many social situations. In financial decision making, for example, the choices would involve profit or loss instead of prison terms. Simulation studies of prisoner's dilemma show that the emergence of cooperation depends on the topology of the underlying social network [1, 7, 31, 45, 47, 52, 53].

However, game theoretical approaches are limited to short-term decision making and do not take into account peer influence arising from deeper social relationships. Another approach to understanding the emergence of cooperation in social networks is to examiner the effects of peer influence.

Cohesion in social groups requires some degree of consensus about identity of the group. Dunbar [17] argued that one way in which speech influenced human evolution was to serve as a form of verbal "grooming" by which human social groups maintained cohesion. As supporting evidence he presented comparisons of group sizes in apes and humans. The "natural" group size for baboons and other apes is about 30–60 individuals. However, Dunbar argued that the greater efficiency of speech as social grooming allowed humans to attain a natural group size of 100–150 individuals.

We tested Dunbar's ideas using simulation experiments to determine whether an entire network could achieve consensus after starting from an arbitrary initial configuration. These models represented social groups as Boolean networks [26, 55]. Each individual was represented as a node in the network having one of two possible opinion states (e.g., 0 or 1, YES or NO). Interactions between individuals were confined to pairs having an existing, fixed relationship defined by the network edges. Experiments consisted of sequences of interactions in which pairs of individuals in disagreement could change their minds.

For random networks, the experiments show that consensus could be achieved in networks up to a maximum size and this maximum size depended on the influence that individuals exerted on one another and on the frequency of interactions. Moreover, the maximum represented a phase change, with consensus being common for smaller groups and extremely rare in larger ones.

4.3 Mechanisms for Managing Social Complexity

The traditional way of dealing with complexity in society is to avoid it. Society has evolved several mechanisms to achieve this. From a network perspective, these mechanisms act in one of two ways: to restrict the range of connections between elements in a social network or to restrict the possible connections between potential behaviour that can occur.

Below we look briefly at the ways these mechanisms are implemented in society. We shall examine some of the problems that arise from these mechanisms later (Section 4.5).

4.3.1 *Divide and Rule*

The first mechanism, which can be termed "divide and rule," involves reducing possible connections by dividing a large, complex problem into smaller, more manageable parts. It is typically seen in the hierarchical partitioning of large organizations. Corporations divide their operations into specialized divisions, such as production, marketing, finance, personnel and so forth. In this way, each division deals with a smaller part of the overall problem of managing the business. In large organizations, there will be further specialization, into departments, sections and so on, each dealing with a narrower, more specialized part of operations.

The same principle is applied in many other aspects of daily life. The design of a house, for instance, divides our daily routine into specialized activities. For instance, there are bedrooms for sleeping, a kitchen for preparing food, a bathroom for washing, and a living room for entertaining. The same principle applies in building a house. Specialist workers install electricity, plumbing, masonry, painting and other features.

The divide and rule principle aims to encapsulate parts of a system, treating them as a closed box. The process of encapsulation (or modularity) can lead to hierarchies of modules, as we saw above for large organizations. However, it often fails. Treating parts of a problem, or a system, as a closed box, implicitly assumes that there are no interactions between agents inside the box with those outside. This is frequently not the case and is a common source of problems, both in hierarchical organizations and in society generally.

Society has many mechanisms for encapsulating issues and reducing complexity. Organizations, for instance, are basic building blocks of modern society. In effect they are social solutions to problems [3]. They encapsulate needs and issues under a single sphere, thus providing vehicles for cooperation, coordination and control of activities. Organizations are often formed in "bottom-up" fashion, when groups of individuals recognize a common need or interest. Alternatively, they may be created "top-down" as large bodies (e.g., governments) respond to particular social problems. In small organizations, control is often direct and informal, driven by individual leaders. In contrast large, distributed organizations usually need to employ bureaucratic control methods, including procedures, rules and standards [18, 19].

4.3.2 *Containing Extremes*

Another class of approaches used by society to reduce complexity can appropriately be termed "containing extremes." This approach occurs where institutions turn problems that are extreme for individuals into "means" for society as a whole. Examples include infrastructure resources such as water supply and electricity and the financial resources provided by banks and insurance. The financial resources, for instance, convert an individual's urgent need for a large amount of money in a short period into constant regular payments over a long period. Likewise ambulance, fire brigades, and police convert the problem of dealing with extreme situations that are urgent, dangerous and often traumatic, for individuals into social services for dealing with emergencies.

More generally, societies use specialization to control extremes. This is the basis for much commerce. Any product or task that individuals need intermittently (i.e., as an extreme measure) creates an opportunity for a specialist to provide a commercial service.

Societies also use mechanisms to avoid extremes in individual behaviour. Social controls aim to ensure that individual behaviour conforms to accepted norms and to avoid extreme actions [50]. These controls include social conventions and beliefs as well as peer pressure and enforcement of laws and regulations.

Industrial controls include standards and protocols. These can be enforced as laws to avoid extreme problems (e.g., health and safety) or by commercial pressure (e.g., conforming to standards to achieve product compatibility and flexibility).

4.4 Mechanisms of Unplanned Social Change

Social changes arise in many ways. However, of greatest concern here are not managed changes that occur by planning, but spontaneous changes that occur unintentionally as self-organization with a social system.

The problem of unexpected outcomes of plans and actions has been a long-running concern of philosophers, economists and other thinkers. In 1936, the American sociologist Richard Merton proposed a *law of unintended consequences*, which holds that any action in a complex system inevitably creates unanticipated outcomes [46]. He suggested five underlying reasons:

- *Ignorance*, because people are never aware of all the factors, issues or processes involved in a highly complex system
- *Error*, because of mistakes, including perpetuation of habitual ways of acting
- *Imperious immediacy of interest*, because people act on the basis of immediate, local priorities or self-interest, rather than the entire group or society
- *Values*, including social, traditional or other values, according to which people act even if the consequences are contrary to best interests
- *Self-fulfilling prophecy*, which occurs where fear of a problem actually creates the problem

Merton's famous study treated the problem chiefly from the perspective of decision-making, especially by governments. Here we treat the problem from the perspective of the processes that produce the problem as they occur in complex networks. Although this includes the problem of poor decision-making, we are more concerned to tease out the role played by processes in complex networks.

In the following subsections, we identify several broad classes of mechanisms that are extremely common in complex social systems and underlie many unplanned events and resulting trends. Each arises from the complexity of social networks and social activity. Although distinct, they are all closely related and often act in unison with one another.

4.4.1 *Failure of Models*

Essentially, all of Merton's five causes are all symptomatic of a common underlying problem: the failure of the models that people apply when making decisions about complex systems. Our use of the term "model failure" here ranges from an individual's mental schemas causing misperceptions to flaws in engineering models of large systems.

Many social problems arise from situations in which people's models fail. Important examples of model failure include false assumptions and "closed box thinking", in which people ignore wider influences on a particular situation. It also includes many kinds of false assumptions. One example is assuming constant conditions in a system that is really changing.

4.4.2 *Cascading Contexts*

The richness of interconnections between social activity and economics, environment, technological, and many other phenomena leads to the problem of cascading contexts. It occurs when events in one context set off chains of events in completely different contexts.

Cascading contexts are often involved in accidents. For instance, a car owner is short of money, so puts off car maintenance. This allows a brake problem to develop, which in turn means that the driver is unable to stop in time when another car suddenly pulls out from a side street.

Such cascades pose a huge challenge in trying to plan or predict the consequences of social changes. Despite the best intentions, despite all the planning and care in the world, plans can go haywire. The social consequences can be both serious and unpredictable.

One of the defining features of models (whether formal or mental) is that they are limited. Of necessity models do not include everything. They represent only those features that are essential to understand a system within a particular context. Contexts can cascade because the real world is more complex than any model people might use in thinking or planning. Events in one context can trigger conditions in another context, setting off a whole new chain of events.

Such problems defeat conventional procedures for model interpretation, such as scenarios and sensitivity analysis, because they arise from conditions outside the model concerned.

Cascading contexts pose a constant problem for decision-making. Later (see Section 4.5) we will see that cascading contexts are often involved in side effects of new technologies.

4.4.3 *Positive Feedback*

Feedback occurs where the outcome of some process "feeds back" to become input to the process again converts local variations into large scale patterns. Well known socioeconomic examples include regional differences in real estate prices, clustering of businesses and success of media personalities.

Positive feedback is a common process in complex systems. It leads to the emergence of large scale patterns from local variations. Positive feedback play a prominent part in many processes that lead to ad hoc sorting. For example, we saw earlier (Section 4.2.1) that positive feedback is responsible for stigmergy, the accumulation effect that leads to self-organization in ant colonies.

A more familiar example is the emergence of fame and popularity. A business, say, achieves good reports from some of its customers. They persuade others to use the business. The more successful the business is seen to be, the more attractive it becomes and hence the more successful. Positive feedback of this kind leads to exponential distributions in popularity. Examples of situations where is occurs include tourist destinations, real estate, and media personalities of all kinds.

4.4.4 *Changes in Network Connectivity*

Changes in institutions or mechanisms that bind society together can lead to chaotic and unpredictable changes social behaviour, which can destabilize a system.

Breakdown of a central service can lead to short-term instability. Loss of power supply or law enforcement, for instance, can lead to anarchy. An infamous example of this was the New York blackout of 1977. A series of power station failures led to New York being without electricity for nearly 24 hours. Emboldened by the darkened streets, normally law-abiding citizens went on an unprecedented rampage of looting and violence: "thousands of otherwise law-abiding citizens joined in what was to become the largest collective theft in history" [48].

Permanent changes can lead to a cascading series of social changes. The Industrial Revolution in Britain, for example, forced people to migrate into cities to find work as increasing automation led to expansion of mills and factories. Old social ties were replaced by new ones, leading to breakdown of traditional structure of the extended family, the evolution of trade unions, and financial changes, including the introduction of poor laws, and the emergence of new financial institutions including friendly societies, cooperative societies and savings banks [54]. Similar processes have continued to occur in more recent times as more and more countries become industrialized (e.g., India, Brazil).

At the end of the twentieth century, the Information Revolution led to massive changes in the way people communicate and interact with each other. An obvious change wrought by mobile phones is that social arrangements are apt to be more fluid. In previous times, people planning to meet would need to specify a time and place well ahead. With the advent of mobile phones, however, they can change their plans at a moment's notice. A dramatic example of how this has affected interpersonal communication was evident during the Paris riots of 2005 [5]. Police found it difficult to react to in time to the disturbances because the rioters did not make plans ahead of time, but initiated their attacks spontaneously. When a small group of rioters found an area that was not patrolled, the word would quickly spread by text messages and large mobs could assemble within minutes.

We shall see examples of the effects wrought by modern communications technology in Section 4.5.3.

4.4.5 *The Brittleness of "Divide and Rule" Strategies*

The strategy of "divide and rule" is the most widespread approach to dealing with complexity (social or otherwise). As we saw earlier, this approach carves up a large problem (or organization) into separate, smaller and therefore simpler ones. Many social institutions and practices seek to reduce and control complexity this way. Large companies, for instance, separate their operations into division, such as production, finance and marketing. Familiar examples of divide and rule include the way homes are divided into rooms with different functions, such as dining, washing or sleeping.

The problem is that divide and rule treats any issue essentially as a closed box, unconnected to other issues or processes. But no system is really closed, so matters can rarely be dealt with, in isolation. The real world is richly connected. The divide and rule strategy fails when situations arise that cut across the subdivisions, or when essential connections are broken. Large-scale operations often run into problems when some activity cuts across different divisions. A major train smash, for example, would require coordination between ambulance and police, as well as the railway company involved, government safety regulators and insurance investigators.

4.5 Side Effects of New Technology

In the long run of human history, one of the most prominent stimuli for social change has been technological change. Tool-making, the control of fire and the introduction of farming heralded changes from hunter-gatherer societies to fixed settlements. The invention of writing arguably played a role in the origins of civilization. In more recent times, the invention of printing accelerated the spread and availability of ideas during the Renaissance. The Industrial Revolution triggered large-scale social change [32].

In the introduction of new technology, we can see at work many of the processes described earlier (Section 4.4). As examples, we examine two cases—the effects of introducing labour-saving devices into family homes during the second half of the twentieth century, and the social consequences of new modes of communication since 1990. In both cases the patterns of social change they initiated are still in progress, but statistical evidence clearly reveals that society has already undergone major changes [28, 29].

4.5.1 *Labour-Saving Devices*

The decades following the end of World War II saw rapid population growth in many countries and a rapid growth in prosperity. Industry responded by adapting a string of inventions for use as labour-saving devices in homes. These inventions included washing machines and dryers, electric irons, refrigerators and freezers, vacuum cleaners, electric sewing machines, electric mixers and frying pans, dish-washers, and microwave cookers.

Many of these devices had been in industrial use for decades, but were not adapted for home use until after World War II. The rate of spread was uneven too. Many devices became available in the United States first, but other western countries soon followed. United Nations figures on sales of electrical goods shows that markets have reached saturation levels in western countries, but continue to spread in many countries even today [58].

All of the goods mentioned above were sold to reduce the time needed for housework and to make life more comfortable. However, providing people with more free time set off a cascade of unexpected socioeconomic side effects [29]. We can summarize just a few of these effects as follows.

1. *Increase in the number of married women in the workforce.* Prior to World War I, less than 10% of married women held a paying job (Fig. 5.5). By the early 21st century the figure exceeded 80%.
2. *Increase in child day care.* With both parents working, the need increased for centres to care for pre-school children, and for after-school care. This need led to a steady increase in day care centres. In Australia the number of centres doubled in the 20-year period 1980–2000 [2].
3. *Increase in the cost of living.* With more money to spend, families were able to spend more on housing. Demand drove prices up. By the early 21st century, two incomes in a family were virtually a necessity.
4. *Increase in divorce rate.* With couples no longer financially dependent on each other, there was no longer any necessity for unhappy couples to remain together.

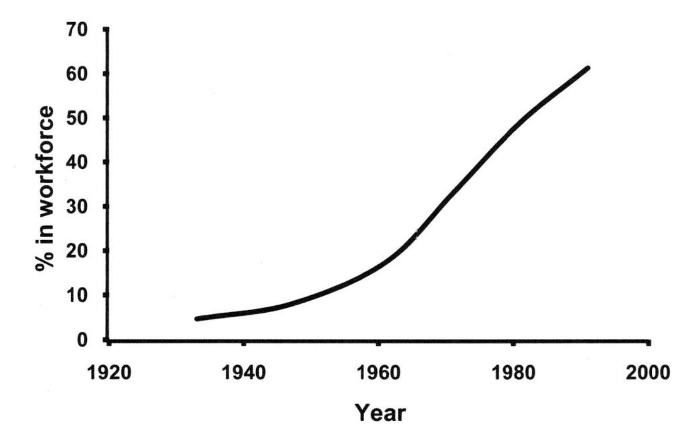

Figure 4.5 Increasing trend in the percentage of married women in Australia aged 25–34 who are in the workforce. Data from the Australian Bureau of Statistics.

5. *The rise of fast-food industries.* Since in many families, all adults were in full-time employment, there was limited time for basic domestic tasks such as cooking.

4.5.2 *Technology and Social Isolation*

The expanding range of technologies in modern society has had many side-effects on everyday social life, especially in western countries. One of these effects is to reduce the frequency of social contact during the course of a normal day.

Among the many social "conveniences" introduced around the turn of the Millennium were automated teller machines (ATMs), Internet and phone banking, automated messaging, self-service petrol, self-serve checkouts and online shopping. All of these automated services, as well as many other innovations, have the effect of reducing the need for social interaction during the course of everyday business activities. As we shall see in the following section, advances in communication have also served to isolate people from one another.

The effect of technology on entertainment has likewise tended to increase social isolation. From radio and television to video and home theatre, it has become increasingly easy to obtain high quality entertainment at home.

Computer games also provide a growing source of solitary entertainment. Multiplayer games have become highly popular, but it remains unclear whether they capture key elements of social interaction. At the same time, addiction to computer games is increasingly recognized as a social problem, exemplified by the young Korean couple who in 2009 allowed their real baby to starve to death while they were caring for a simulated child [13]. As we saw earlier, there is also evidence that media (if not social separation alone) has contributed to increasingly violent behaviour in society [62].

Putnam [50] drew awareness to the decline in participation in community activities that had taken place across America since the 1950s. On the other hand, critics argue that social activities have merely been transformed from community activities to the workplace.

Nevertheless there is evidence that social isolation is real and that it has negative effects on society. Perhaps the most startling evidence comes from cases of people dying alone and being undetected for long periods (e.g., [4]). Perhaps the most extreme case was that of a London woman who lay dead in her apartment with presents under a Christmas tree and the TV running for more than two years before being discovered [21].

4.5.3 *Information and Communications*

The late decades of the 20th century saw a revolution in information and communications technology. Within the space of 10 years, the spread of mobile phones, text messaging, and Internet technologies, such as email, weblogs, webcams, Internet forums, instant messaging and online social networks, have dramatically altered communication patterns throughout the world (Fig. 4.6). Cheap digital still and video cameras provide individuals with an unprecedented capacity to document their worlds, while free online publishing enables creative works to be disseminated instantly. This revolution has already had important side-effects on social behaviour and organization.

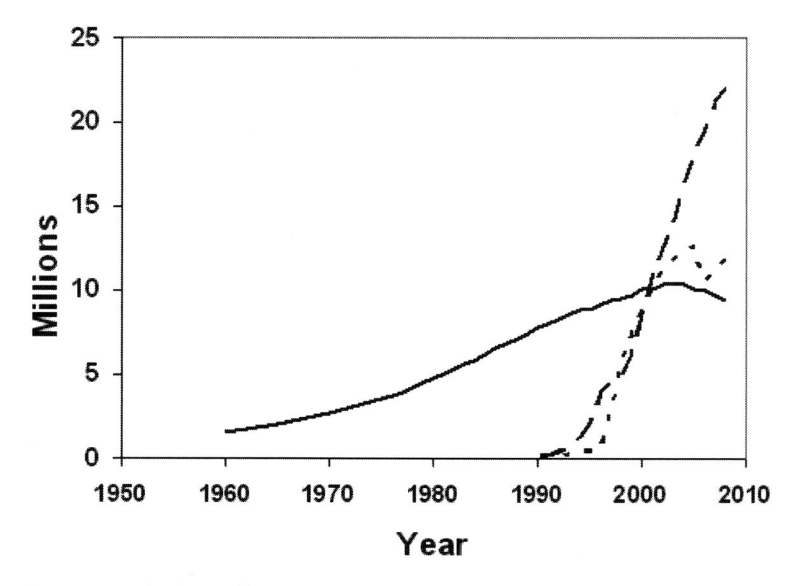

Figure 4.6 A phase change in communication occurred during the period 1995–2005. Here figures from the Australian Bureau of Statistics over the 50 year period 1960–2010 reveal the change for Australia. At the start of the period most communication was by fixed phone lines (solid line). However the number of these connections declined as the number of mobile phones (dashed line) and internet accounts (dotted line) increased in Australia [59].

Each of the technologies described above facilitates communication, enabling people to freely broadcast experiences and ideas efficiently to an unlimited audience. Before the advent of the Internet, such communicative power was confined to large organizations and celebrities: media companies who owned television networks, newspapers and publishing houses, the individuals these companies deemed newsworthy, and companies who could afford to buy advertising space. Today, a message, photo or video can be created by virtually anyone and distributed worldwide in seconds. Thus the Internet undeniably presents an enormous leap in the democratization of information. The rapid dissemination of information, along with advances in information analysis tools, undoubtedly underlies the continuing acceleration in technological development seen in many fields. Growth in independent publishing also has the potential to foster human creativity by allowing artists and authors to control the publication process and claim the full profit from their work; a number of mainstream authors have recently begun self-publishing for this reason [15, 41].

A key factor in the success of the Internet is the realization that crowd behaviour, in appropriate conditions, will self-regulate to organize useful information. Search engines rank websites by counting incoming links from other sites. The Web encyclopaedia Wikipedia, which consists entirely of content written by users, now approaches the *Encyclopaedia Britannica* in accuracy and far exceeds it in scope [20]. Retail and auction websites allow users to rate and review products and sellers, creating reputation systems that efficiently punish cheats without active policing, and thus allow relatively safe transactions between strangers who may be continents apart [37]. The abundance of social information also enables more sinister applications. Recently, a research group has claimed to predict the direction of daily stock market fluctuations with over 80% accuracy by analysing mood-related words posted on Twitter [10]. More generally, it seems clear that large-scale data harvesting provides unprecedented opportunities for market analysis and manipulation.

The human consequences of this change are even harder to assess. In a recent book, Nicholas Carr argues that the Internet is restructuring human cognitive function by rewarding shallow, rapid information processing at the expense of serious contemplation [11].

Ironically, advances in communication technologies may also work to increase social isolation. Broadcast communications are not as conducive to strong personal relationships as private ones. Empathic responses and relationship formation involve numerous subtle non-verbal cues and exquisitely timed interaction patterns [16] that have no analogue in the online world. Text, voice and even video lack many of the cues inherent in natural primate communication. A recent study found that people felt increased loneliness after chatting online compared with face-to-face chatting [35].

As discussed above, the need for direct human interaction in day-to-day life has been virtually eliminated due to automation of ordinary activities. Up until the mid-20th century social interactions might be characterized as geographically based. Our ancestors lived in tribes where relationships were mostly lifelong and arose through continual, largely involuntary contact resulting from group living. The advent of large cities brought increased mobility and contact with strangers, but social groups and activities were still largely confined to local neighbourhoods. In contrast, social interaction today is almost entirely voluntary and to a large extent independent of geography. The resultant relationships cluster around common interests, but it is unclear whether they provide the same qualities as relationships of the past.

For individuals unable to adapt to this new environment—particularly the elderly and socially or intellectually impaired—the consequence may be severe social isolation. Junghyun et al. [38] found that lonely individuals were susceptible to compulsive Internet use, which could further isolate them. Robots have been developed to address the social needs of the elderly and appear to provide considerable benefits [39]. However, many argue that the unwillingness of the younger generation to personally care for their elders is symptomatic of a growing shallowness in modern relationships.

At the same time, the ubiquity of information technology provides unlimited access to a wide range of supernormal stimuli and allows individuals to vividly simulate activities that would normally attract social censure (such as theft, rape and murder). Although the issue is contentious, various studies support the view that viewing and simulated participation in violence increases antisocial behaviour in real life [36].

Evidence for direct negative psychological effects of Internet use, however, remains ambiguous. An early study suggested that Internet adoption was associated with decreased well-being and weaker relationships, but later examination of the same study group found that this effect dissipated over time. Recent meta-analysis found only a very weak negative relationship between Internet use and measures of well-being such as depression, loneliness, self-esteem and life satisfaction [12]. Several studies suggest that the Internet improves or is neutral for the well-being of most users, but harmful for a susceptible minority [63].

Nonetheless, some disturbing patterns are apparent. One study found a 40% decrease in empathic response among college undergraduates within the last 20 years, with the sharpest drop occurring in the last decade [42]. Research across several mammal species shows that appropriate socialization is crucial to normal psychological development, with inadequately socialized individuals exhibiting a wide range of abnormal responses. Perhaps weakened human relationships, combined with abundant pathogenic stimuli, often fail to provide the needed context for healthy psychological development.

At a broader level, the Internet has been hailed as a potential vector for social and political change. Freely available broadcast communications enable grassroots movements to rapidly recruit and mobilize supporters in response to current events. Theoretically, this could make these movements more effective. For example, commentators hailed the role of the microblogging service Twitter in the 2010 Iranian elections. However, some argue that the ease of participation in Internet campaigns actually decreases their efficacy [22]. Internet-based social movements allow individuals to feel—and advertise—that they are participating in a cause without offering any meaningful support. Gladwell noted that over 1 million participants in a recent Facebook campaign to "Save Darfur" have donated an average of 9 cents each. Significant commitments to social causes require personal ties to the movement; the weak ties provided by social media, Gladwell argues, do not provide this. In addition, studies show that when individuals feel they are already behaving virtuously, their inclination to further acts of altruism declines; perhaps individuals who affiliate themselves with Facebook campaigns are less likely to bother with more significant action.

4.6 Conclusion

The complexity of modern society poses many challenges for individuals and governments alike. As the above discussion and examples have shown, complexity in human society is a source of many unplanned trends and many unexpected problems. A particular challenge that we have explored is the problem of coping with the rapid and unexpected social side-effects that arise from the introduction of new technologies. Other challenges include the tremendously interconnected nature of today's economies and the complex inter-relationships between society and the environment.

Our central message is that a number of processes arising from the complexity of social systems, underlie a great many unexpected trends and events in society. As we have seen, the very nature of these processes makes prediction difficult.

Perhaps the most difficult to deal with is the problem of cascading contexts, since it highlights the difficulty of developing models that are adequate to identify the full range of issues that may arise in highly complex systems. The process of cascading contexts implies that seemingly small changes in one domain can vastly alter system functions on a broader scale. For modern societies in particular, this observation means that the many technological developments of the past century have had undreamt-of ramifications for social structures and human relationships. As technological change continues to accelerate, we urgently need a better understanding of how the resultant ongoing transformations in social networks feed into human psychology, behaviour and wellbeing—and how social change, in turn, will affect future technologies.

Because many of the problems described above arise through the failure of simplistic models, complex simulations are likely to play a large part in addressing them. We envisage two major roles for simulation.

First, we can use simulations to better understand the network processes that drive social change, such as the spread of ideas and practices, the influence of media and communication systems, and the effects of new technology on social network structure. For these purposes, agent-based simulations, which directly simulate the interactions and choices of individuals in a network, can provide substantial insights.

Second, interactive simulations allow us to explore how people are likely to respond to new social situations and technologies. Just as online computer games provide novel social contexts for human interactions, research simulations using immersive environments provide opportunities to study the human consequences of technological innovation.

Finally, we argue that just being aware of the kinds of processes we have described can help to reduce their impact as we attempt to manage social complexity. As we have tried to show, the consequences of failing to understand complexity, and to consider its effects in planning and decision making are potentially catastrophic.

References

1. Abramson, G., and Kuperman, M. (2001). Social games in a social network. *Phys. Rev. E* **63**, 030901.
2. ABS (2009). *Australian Social Trends*, Australian Bureau of Statistics. http://www.abs.gov.au.
3. Aldrich, H. E., and Marsden, P. V. (1992). Complex organizations, in *Encyclopedia of Sociology*, Vol. 1, ed. Borgatta, E. F., and Borgatta, M. L. (MacMillan, New York), pp. 271–277.
4. Anonymous (2006). Dying alone. *The Daily Telegraph*, Sydney, February 22.
5. Anonymous (2009). 2005 civil unrest in France, *Wikipedia*. en.wikipedia.org/wiki/2005_civil_unrest_in_France
6. Axelrod, R. (1984). *The Evolution of Cooperation* (Basic Books, New York).
7. Axelrod, R., and Dion, D. (1988). The further evolution of cooperation. *Science* **242**, pp. 1385–1390.
8. Berk, R. A. (1974). A gaming approach to crowd behavior. *Am. Sociol. Rev.* **39**, pp. 355–375.
9. Blumer, H. (1957). Collective behavior, in *New Outline of the Principles of Sociology*, ed. Lee, A. M. (Barnes and Noble, New York), pp. 166–222.
10. Bollen, J, Huin, M., and Zeng, X. (2010). Twitter mood predicts the stock market, arXiv:1010.3003v1
11. Carr, N. (2010). *The Shallows: What the Internet Is Doing to Our Brains* (W. W. Norton, New York).
12. Huang, C. (2010). Internet use and psychological well-being: a meta-analysis, *Cyberpsychol. Behav. Soc. Networking* **13**, pp. 241–249. doi:10.1089/cyber.2009.0217.

13. Cho, J. (2010). Game addicts arrested for starving baby to death. http://abcnews.go.com/International/TheLaw/baby-death-alleged-result-parents-online-games-addiction/story?id=10007040
14. Dawes, R. M. (1980). Social dilemma. *Ann. Rev. Psychol.* **31**, pp. 169–193.
15. Deahl, R. (2010). Agents weigh the growth of alternate publishing options. *Publisher's Weekly*, May 24.
16. De Waal, F. (2009). *The Age of Empathy: Nature's Lessons for a Kinder Society* (Crown Publishing Group, New York).
17. Dunbar, R. (1996). *Grooming, Gossip and the Evolution of Language* (Faber and Faber, London).
18. Edwards, R. E. (1979). *Contested Terrain: The Transformation of the Workplace in the Twentieth Century* (Basic Books, New York).
19. Fligstein, N. (1991). *The Transformation of Corporate Control* (Harvard University Press, Cambridge, MA).
20. Giles, J. (2005). Internet encyclopaedias go head to head. *Nature* **438**, pp. 900–901
21. Gillan, A. (2006). Body of woman, 40, lay unmissed in flat for more than two years. *The Guardian*, April 14. http://www.guardian.co.uk/uk/2006/apr/14/audreygillan.uknews2
22. Gladwell, M. (2010). Small change: why the revolution will not be tweeted. *The New Yorker*, October 4.
23. Granovetter, M. (1978). Threshold models of collective behaviour. *Am. J. Sociol.* **83**, pp. 1420–1443.
24. Green, D. (1994). Emergent behaviour in biological systems. *Complexity Int.* **1**, green01. www.complexity.org.au/vol01/green01/
25. Green, D.G. (2000). Self-organization in complex systems, in *Complex Systems*, ed. Bossomaier, T. J., and Green, D. G. (Cambridge University Press, New York), pp. 7–41.
26. Green, D. G., Leishman, T. G., and Sadedin, S. (2006). The emergence of social consensus in simulation studies with Boolean networks, in *Proceedings of the First World Congress on Social Simulation*, Kyoto, Vol. 2, in Takahashi, S., Sallach, D., and Rouchier, J., pp. 1–8.
27. Green, D. G. (2010). Elements of a network theory of complex adaptive systems. *Int. J. Bio-Inspired Comput.* **3**(3), pp. 159–167.
28. Green, D. G. (2012). *Of Ants and Men: The Struggle to Cope in a Complex World* (Springer, Berlin).
29. Green, D. G., Leishman, T. G., Sadedin, S., and Leishman, G. D. (2010). Of ants and men: the role of complexity in social change. *Evol. Inst. Econ. Rev.* **6**(2), pp. 41–57.

30. Hardin, R. (1992). *Organizational Trust* (Oxford University Press, Oxford).
31. Harrald, P. G., and Fogel, D. B. (1996). Evolving continuous behaviours in the iterated prisoner's dilemma. *BioSystems* **37**, pp. 135–145.
32. Hartwell, R. M. (1971). *The Industrial Revolution and Economic Growth* (Methuen, London), pp. 339–341.
33. Hogeweg, P., and Hesper, B. (1983). The ontogeny of the interaction structure in bumble bee colonies: a MIRROR model. *Behav. Ecol. Sociobiol.* **12**, pp. 271–283.
34. Hogeweg, P., and Hesper, B. (1990). Individual-oriented modelling in ecology. *Math. Comput. Modell.* **13**(6), pp. 83–90.
35. Hu, M., and McDonald, D. (2008). Social Internet use, trait loneliness, and mood loneliness. Paper presented at the annual meeting of the International Communication Association, TBA, Montreal, Quebec, Canada, May 21. 16 http://www.allacademic.com/meta/p234713_index.html
36. Huesmann, L. R. (2007). The impact of electronic media violence: scientific theory and research. *J. Adolescent Health* **6**, pp. S6–S13.
37. Jøsang, A., Ismail, R., and Boyd, C. (2005). A survey of trust and reputation systems for online service provision. *Decision Support Sys.* **43**, pp. 618–644
38. Kim, J., LaRose, R., and Peng, W. (2009). *CyberPsychol. Behav.* **12**(4), pp. 451–455. doi:10.1089/cpb.2008.0327.
39. Kidd, C. D., Taggart, W., and Turkle, S. (2006). A sociable robot to encourage social interaction among the elderly. *Proceedings, International Conference on Robotics and Automation ICRA 2006*, pp. 3972–3976.
40. Klüpfel, H., Schreckenberg, M., and Meyer-König, T. (2005). Models for crowd movement and Egress simulation, in *Traffic and Granular Flow*, ed. Hoogendoorn, S. P, Luding, S., Bovy, P. H. L., Schreckenberg, M., and Wolf, D. E. (Springer, Berlin). doi: 10.1007/3-540-28091-X
41. Konrath, J. A. (2010). eBooks and the ease of self-publishing. *The Huffington Post*, October 16. http://www.huffingtonpost.com/ja-konrath/ebooks-and-self-publishing_b_764516.html
42. Konrath, S. H., O'Brien, E. H., and Hsing, C. (2010). Changes in dispositional empathy in American college students over time: a meta-analysis. *Personality Social Psychol. Rev.* doi: 10.1177/1088868310377395.
43. Kuper, A., and Kuper, J. (1996). *The Social Science Encyclopedia* (Routledge, New York).
44. Leishman, T. G., Green, D. G., and Driver, S. (2009). Self-organization in simulated social networks, in *Computer Mediated Social Networking*

LNAI5322, ed. Purvis, M., and Savarimuthu, B. T. R. (Springer, Berlin), pp. 150–156.

45. Lindgren, K., and Nordahl, M. G. (1994). Evolutionary dynamics of spatial games. *Phys. D* **75**, pp. 292–309.
46. Merton, R. K. (1936). The unanticipated consequences of purposive social action. *Am. Sociol. Rev.* **1**(6), pp. 894–904.
47. Nowak, M.A., and May, R.M. (1992). Evolutionary games and spatial chaos. *Nature* **359**, pp. 826–829.
48. NYB (2000). Eyewitness report, *The Blackout History Project*, George Mason University. http://www.blackout.gmu.edu/
49. Orbell, J. M., and Dawes, R. M. (1993). Social welfare, cooperators advantage and the option of not playing the game. *Am. Sociol. Rev.* **58**, pp. 787–800.
50. Putnam, R. (2001). *Bowling Alone: The Collapse and Revival of American Community* (Simon & Schuster, New York).
51. Reynolds, C. (1987). Flocks, herds and schools: a distributed behavioral model. *ACM SIGGRAPH Comput. Graphics* **21**(4), pp. 25–34.
52. Ridgeway, C. L. (1988). Compliance and conformity, in *Encyclopedia of Sociology*, Vol. 1, ed. Borgatta, E. F., and Borgatta, M. L. (Macmillan, New York), pp. 277–282.
53. Schweitzer, F., Behera, L., and Mühlenbein, H. (2002). Evolution of cooperation in a spatial prisoner's dilemma. *Adv. Complex Sys.* **5**, pp. 269–299.
54. Smelser, N. (1959). *Social Change in the Industrial Revolution: An Application of Theory to the British Cotton Industry* (The University of Chicago Press, Chicago).
55. Stocker, R., Green, D. G., and Newth, D. (2001). Consensus and cohesion in simulated social networks. *J. Artif. Soc. Social Simul.* **4**(4). http://www.soc.surrey.ac.uk/JASSS/4/4/5.html
56. Stocker, R., Cornforth, D., and Green, D. G. (2003). A simulation of the impact of media on social cohesion. *Adv. Complex Sys.*, **6**(3), pp. 349–359.
57. Turner, R. H., and Killian, L. M. (1993). *Collective Behavior*, 4th ed. (Prentice-Hall, Englewood Cliffs, NJ).
58. UNSD (2009). *Industrial Commodity Statistics Database*, United Nations Statistics Division. http://data.un.org/
59. United Nations (2010). *UNData*. http://data.un.org/
60. Waddington, D., Jones, K., and Critcher, C. (1989). *Flashpoints: Studies in Public Disorder* (Routledge, London).

61. Wikipedia (2005). 2005 civil unrest in France. en.wikipedia.org/wiki/2005_civil_unrest_in_France
62. Williams, T. M. (1986). *The Impact of Television: A Natural Experiment in Three Communities* (Academic Press, New York).
63. Yu, S., and Chou, C. (2009). Does authentic happiness exist in cyberspace? Implications for understanding and guiding college students' Internet attitudes and behaviours. *Br. J. Educational Technol.* **40**, pp. 1135–1138.

Chapter 5

Developing Agent-Based Models of Business Relations and Networks

Ian Wilkinson,[a] Fabian Held,[a] Robert Marks,[b] and Louise Young[c]

[a] *Discipline of Marketing, The University of Sydney Business School, University of Sydney, Sydney, New South Wales 2006, Australia*
[b] *School of Economics, Australian School of Business, University of New South Wales, Sydney, New South Wales 2052, Australia*
[c] *School of Marketing, College of Business and Law, University of Western Sydney, Penrith, New South Wales 2751, Australia*
ian.wilkinson@sydney.edu.au, fabian.p.held@gmail.com, r.marks@unsw.edu.au, louise.young@uws.edu.au

Business relations and networks play a central role in the way business and economic systems are organized and function. But their dynamics and evolution have received limited research attention, with research focusing on comparative static and cross-section surveys. In order to develop appropriate research-based management and policy advice, we need a better understanding of

This chapter is based in part on papers presented in 2010 at the World Congress on Social Simulation (Kassel, Germany), Industrial Marketing and Purchasing (IMP) Group Annual Conference (Budapest, Hungary) and the Australia New Zealand Marketing Academy Annual Conference (Christchurch, New Zealand). The research is supported by an Australian Research Council Discovery Project Grant DP0881799 (2008–2010).

Networks in Society: Links and Language
Edited by Robert Stocker and Terry Bossomaier

ISBN 978-981-4316-28-6 (Hardcover), 978-981-4364-82-9 (eBook)
www.panstanford.com

how business relations and networks form and evolve. One way to do this is through the development of agent-based simulation models of business relations and networks that allow researchers to explore systematically the nature and impact of various factors on the structure, behaviour and performance that are beyond traditional closed-form mathematical solutions and that would be impossible to implement in the field. We identify the main mechanisms driving the evolution of business relations and networks and review models from various disciplines that attempt to represent these. This provides the basis for developing a generic simulation platform called the Business Network Agent-Based Modelling System (BNAS).

5.1 Introduction

Social and business networks are ubiquitous and an increasingly important area of research attention in many disciplines [18, 29, 63]. One of the major challenges is to better understand, predict and control their dynamics, including how they form and evolve and how this shapes their behaviour and performance [88, 95].

Business networks refer to the interdependent systems of intra and interorganizational relations that are involved in markets, including firms, government agencies and other types of organizations [113]. They are increasingly understood as examples of complex adaptive systems (CAS) in which order arises in a bottom-up self-organizing manner, based on the micro interactions taking place over time among the networks, people and organizations involved [24, 72, 77, 113]. In addition, large-scale or macro order has feedback effects on lower level, micro interactions, and the network operates in the context of other connected networks and the broader environment. CAS theory challenges traditional notions of management, policy making and research methods because control is decentralized not centralized and because CAS are highly nonlinear systems exhibiting emergent behaviour and structure.

These characteristics of business networks are more clearly recognized and appreciated, as economies become increasingly internationalized, interconnected and interdependent [57] as a result of revolutions in commerce and communication technologies that broaden, deepen and speed up the processes of interaction among economic actors around the world. Moreover, we recognize networks and their effects more clearly because network theories and research methods have developed rapidly over recent years, driven by developments in the science of networks and advances in computer technology.

The increasing attention being given to business networks is apparent in all business disciplines, in which the nature, development and performance of firms, markets, regions and nations are increasingly being linked to the way economic actors are interconnected, rather than being attributed solely to their individual characteristics. In marketing, this focus can be traced to research in channels and B2B markets in the United States, Europe and Australasia [112] and is of continuing importance [49, 113, 115]. In management, strategy and economics, the topic increasingly features in books, articles and special issues in leading management and economic journals [e.g., 29, 63, 84, 88].

The challenge for managers and policy makers is not about the management and control of such systems but how to participate and manage in them [90]. Government agencies and policy makers do not operate outside business networks: instead, they are part of them [107], along with the other people, firms and organizations. All participants are to some degree interdependent; it cannot be avoided. No one is in overall control, though some may have far more influence on parts of the system than do others. Overall behaviour and performance is not a simple sum of the behaviour of the actors involved: what happens depends on the ways different actors behave and respond and the direct and indirect interactions among these actions and responses taking place over time.

In order to provide research-based advice to managers and policy makers, we need to understand the dynamics of business relations and networks, including how they form and evolve. Unfortunately, past research is dominated by linear, comparative-static, variable-based theories and methods, in which time and process are mostly absent. Structure, behaviour and context are summarized in terms of models of fixed entities (people, firms, relations, networks) with variable properties [1, 86] tested mainly via cross-sectional surveys of managers' reports and perceptions

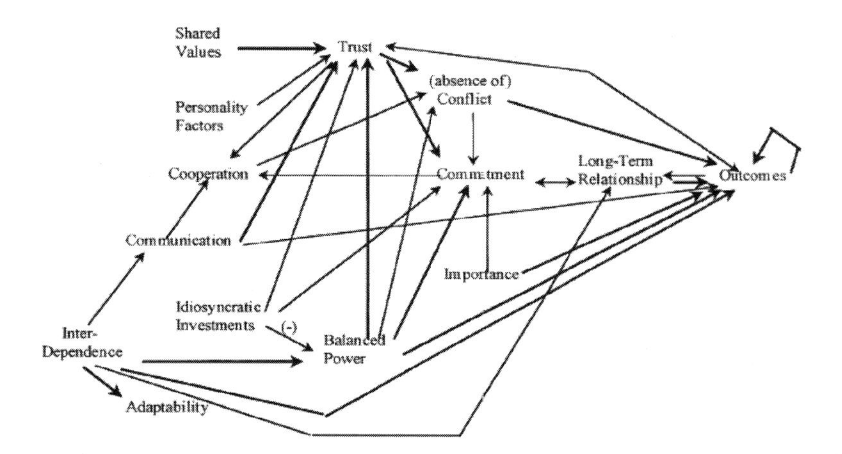

Figure 5.1 Links between dimensions of relations. Reprinted from Ref. 62, copyright Elsevier (1999).

designed to measure the relevant variables and the patterns of covariance. A summary of the main variables and their posited links is shown in Fig. 5.1.

Such research is of limited value in showing how real systems behave and evolve over time because it provides no understanding, tests or observations of the underlying causal mechanisms driving these. Variables do not exist and act in business systems: they are research abstractions produced by operational processes. Driving mechanisms, for example, the forces that enable survival and growth, may be alluded to in developing models and hypotheses but are not directly identified, modelled or tested.

There has been some previous research concerning the dynamics and evolution of business relations and networks [112]. This includes

- stage models of business relations, in which a pre-given sequence of stages is assumed to occur, with each providing the preconditions for the next, allowing analysis of the processes involved in relation and network development [28, 40]

- case studies including descriptive characterizations of relationship and network histories that highlight some of the processes occurring [e.g., 6, 59, 66, 81, 91]
- partial theories and schematic models suggesting some of mechanisms and feedback effects that drive the dynamics of aspects of relations, such as trust and power [e.g., 6, 59, 60, 81, 91, 111]
- speculative descriptions characterizing patterns of change and evolution [e.g., 50]
- description and analysis of some mechanisms and processes involved in the dynamics and evolution of business relations and networks [e.g., 48, 52, 99, 111, 116]
- general theories of organizational change and development that are relevant to business relations and networks [e.g., 3, 104]

5.2 The Role of Simulation and Agent-Based Models

One way of advancing research is through the use of computer simulations, especially agent-based methodologies, which are becoming ever more widely appreciated and used in science [45], including in marketing [79, 106] and economics [101]. Simulation represents a third way of doing science in contrast to induction and deduction [8]. Its uses are manyfold, including prediction, performing tasks, training, entertaining, educating, existence proofs and discovery [8]. For scientific research, the main goals are prediction, existence proofs and discovery.

One example of a successful discovery about the behaviour of networks achieved through simulation is [106] which challenges theories about the importance of critical individuals (opinion leaders) in diffusion processes. These simulations suggest that under most conditions, large cascades of diffusion across social networks are driven not by "influentials" but by a critical mass of easily influenced individuals. In training and education, simulations have the potential to present abstract theories in a visually appealing and engaging way. For example, a simulation of the "Beer Game" can give managers and business students a feel of the many interdependencies in supply chains [58], and agent-based models (ABMs) lying behind most of the online and computer games that have become popular, such as Second Life, Simlife, SimCity and The Sims. Along similar lines, "flight simulators" can be developed to allow managers and policy makers to experiment with alternative strategies in complex systems that are beyond analytical solutions, individual control or simple rules of thumb.

There has been previous research using simulation methods to study business relations and networks. Starting in the 1960s, some computer simulations of business systems were developed, including Forrester's models of industrial dynamics [41], models of market processes [13] and logistics simulation models [19]. These models were limited by the lack of accessible computing power and suitable programming languages. They are all examples of System Dynamics models, in which the dynamics depends on starting conditions and the fixed structure imposed in the form of a set of differential equations linking variables related to representative actors. They allow neither structures to change over time or in response to behaviour, learning and outcomes, nor emergence.

More recent simulations have used Boolean rules to represent the way exchanges are connected [30, 32, 114] and to represent aspects of the evolution of a particular industry [39]. But only limited attempts have been made to develop more comprehensive simulation models that can be calibrated and tested against the dynamics and known histories of real relations and networks.

5.3 Agent-Based Models

ABMs represent a revolution in the way scientists can build models to replicate, analyse, test and predict the behaviour of CAS, because they are able to go beyond the heroic simplifications and assumptions required to solve other types of models. Their relevance and importance is being increasingly recognized in many disciplines, though less so in marketing and business. The inadequacies of current models have been highlighted by to the failure of traditional economic models to anticipate the global financial crisis, leading researchers, government and business to focus attention on building more realistic ABM of socioeconomic systems [e.g., 14].

An ABM is a model formalized in computer code that represents a collection of mathematical rules, applied to a clearly defined set of inputs [36, 45]. These rules transform the inputs into outputs, and, through systematic computation and analysis of the space of possible input values, a mapping of the respective outcomes can be calculated. In contrast to many mathematical models, simulation models need not be so strongly simplified that they become analytically tractable. ABMs allow us to build, investigate and test models of complex systems that are beyond the reach of traditional analytical methods, though they may be summarized in algebraic form [73]. They are not restricted to general statistical models of behaviour, central driving equations or representative agents, but can represent in more realistic ways the microinteractions taking place and how they produce macrostructures and patterns of behaviour. They are not based on a set of general driving equations, as in System Dynamics simulations, but allow for bottom-up control, via a heterogeneous set of interacting agents that can represent people and firms, as well as other actors and objects such as markets, environments and resources, each with its own characteristics, predispositions and rules of behaviour. These rules of behaviour can change as a result of learning via interaction and feedback [101]. Statistical and mathematical analysis and variable-based accounts of system behaviour are still relevant: they play an essential role in developing and testing ABM and in summarizing their output, as they do for real-world social and economic systems.

ABMs offer a middle ground between "thick" and "thin" descriptions [78]. Thick descriptions result from in-depth case studies of actual histories, which reveal the complex causal processes involved but cannot be easily generalized. Thin descriptions result from sample, survey-type research that is more generalizable but abstracts away from any meaningful examination of the processes, events or choices by which different types of variables are interrelated and affect outcomes. Between these two extremes, ABM simulations have been characterized as "opaque thought experiments" [25].

Instead of being limited to a study of what has happened, ABMs allow us to test counterfactuals and to synthesize new forms of behaviour and organization that have not existed in real life, which is why this approach is sometimes referred to as Artificial Life [2, 71]. "[Synthesis] extends the empirical database upon which the theory of the discipline is built beyond the often highly accidental set of entities that nature happened to leave around for us to study" [71 p. ix]. We cannot go back and rerun history to see how sensitive the outcomes that did emerge are to different factors and interventions, but ABMs enable us to capture important features of the process, identify "tipping points" and conduct computer experiments of the impact of different factors [68]. Different theories or assumptions about a system can be implemented and examined.

5.4 Explaining Dynamics: Identifying and Modelling Causal Mechanisms

In order to develop realistic, comprehensive models of the dynamics and evolution business networks that can incorporate and build on past descriptions, characterizations and partial theories, we need to focus on the underlying mechanisms and processes driving change, not on variables and their covariances. Mechanisms are the "verbs" of social and business life, and provide greater insight than do the "nouns" and "adjectives" of variable-based research. The term "mechanism" is used loosely and can be confusing. It has to do with how something has an effect: the actual physical, mechanical, psychological, economic, sociological and so on processes that are involved in one thing leading to another: "a real process in a concrete system, such that it is capable of bringing about, or preventing, some change in the system as a whole" [21 p. 414]. More comprehensively: "Mechanisms consist of entities with their properties and the activities that these entities engage in, either by themselves or in concert with other entities ... a constellation of entities and activities that are organized such that they regularly bring about a particular type of outcome" [53]. Managers and policy makers work with mechanisms and are players in various types of mechanisms driving business networks. Managers are actors in systems and operate in terms of the sense-making processes and actions available to them and the consequences these have for themselves and others, but variables are not actors, rather they are abstractions of the process occurring.

A focus on mechanisms and processes leads to a different approach to explanation that is more relevant for understanding dynamics and change. As summarized by Herbert Simon, "to 'explain' an empirical regularity is to discover a set of simple mechanisms that would produce the former in any system governed by the later" [7]. Mechanisms are often left implicit in our causal explanations, especially if we are focusing on the behaviour of variables rather than actors and events [22, 103]. Thus, we may say that a manager's trust in another firm depends on how reliably or benevolently a firm has acted in the past. This implies that if the other firm's behaviour in terms of reliability and benevolence changes, then it could affect the amount of trust the manager has in it. But left unspecified are the processes involved: what kinds of changes in behaviour are perceived by the manager as evidence of better behaviour? How is this communicated to and understood by the manager? How likely are accurate perceptions to be due to thresholds of awareness and attention, self-interest, bounded rationality and selective perception processes? How do changes in perception cause changes in trust? Is it a simple addition of responses to events as suggested in some models of behaviour, a more complex psychological sense-making process or is it idiosyncratic to the manager? What happens as a result of changes in the amount of trust, and how and why? Such questions cannot be answered by a variable-based approach, although this remains relevant as a way of validating and testing ABM of mechanisms.

Mechanisms shape and influence much of our social world, but identifying them is not something social scientists are trained to do. Natural scientists and engineers have been much more concerned with mechanisms, as they try to understand how things work and use this knowledge to make other things. There the focus is on synthesis rather than analysis. Analysis functions by breaking up existing systems into parts and subsystems in order to understand how they work. Synthesis focuses attention on explaining by growing or replicating some pattern of behaviour or outcomes; what Epstein (2006) refers to as "generative social science" [35].

Synthesis is the basis for major advances in science. This is reflected in developments in chemistry in the 1850s with the compilation of the table of the elements and discovery of how to make elements that did not exist in nature and a century later in the revolution that took place in biology after the discovery of the structure of DNA. Experimentation in social and economic systems is more complex because people have expectations about the future and the rules of behaviour are not fixed but change and evolve. Moreover, these systems are less amenable to systematic experimentation. But developments in computing power and ABM have opened up a new way to proceed. Similar revolutionary advances can be expected in business and social science, as we are now able to build ever-more sophisticated ABM of the many psychological, social and economic mechanisms, demonstrating how they interact over time to produce the business and social systems we observe. We can then use the same models to explore what could happen–economic and business life as it could be–or what is sometimes referred to as "Artificial Life" [71].

5.5 Mechanisms Driving Business Relations and Networks

A comprehensive cross-disciplinary review of previous research regarding business relations and networks has resulted in the identification of five broad types of mechanisms and processes, as well as numerous examples of each type. These are summarized in columns 1 and 2 of Table 5.1. In addition, we have reviewed previous research in which one or more of the mechanisms identified was modelled in a precise way: in the form of computer code, mathematical formulae or other precise specification. The review covered disciplines, including biology, ecology, physics, engineering, artificial intelligence and the social, economic and business sciences [54, 55]. Examples were found for most but not all mechanisms and are illustrated in column 3 of Table 5.1.

Table 5.1 Business relations and network mechanisms and models

Types	Example of mechanisms	Example of models
1. Business acting and specializing	Producing, consuming, buying, selling, learning, copying, targeting, in-sourcing, outsourcing, innovating, firm creation and demise, innovation, opportunity discovery/enaction	Economic models of scale and scope efficiency, transaction cost and production functions [26] Choice and evaluation models [80] Imitation and learning models [20] Specialization [33, 93] Central place and gravity models [4, 23])
2. Business mating	Finding, being found, attracting, homophily, repelling, choosing, being chosen	Trade and Network formation models [16, 65, 106] Attraction and Gravity models [4, 23, 61, 94] Matching, alliance models [43] Partner and trade choice models [12, 74, 98, 109]
3. Business dancing	Interacting, exchanging, cooperating, defecting, responding, initiating, trusting, liking, committing, learning, adapting, terminating	Iterated game models [9, 100, 117] Bargaining models [65, 67] Trust models [64, 102] Evolution of cooperation models [56, 70, 85] Learning models [20] Attraction and loyalty models [67]
4. Interconnecting	Enabling and constraining effects of other relations, comparing, accessing, prioritizing	Learning models [20, 97] Competition and trade models [10, 97] Diffusion models [46, 96] Recommendation engines (e.g. Amazon) Network Co-Evolution models [34, 42, 47]
5. Environmental impact	History, enabling and constraining effects of exogenous environment	Starting conditions Parameter settings Location mapping Environment processes [11]

5.5.1 *Acting and Specializing*

The work of business comprises many types of interrelated activities that are divided up among those involved, and markets and other forms of exchange are used to access specialists' resources and outputs. This forms the basis of all economic and business systems. Firms specialize in particular assortments of activities they perform, which depend on their capabilities and knowledge and the decisions they make, such as about what to outsource or not. Outsourcing decisions involve comparing and evaluating alternative exchange possibilities, which can change as a result of variations in offers and capabilities and the mutual learning taking place. For example, scale and scope, learning and innovation affect costs and performance, which alter outsourcing opportunities and decisions. Opportunities arise for new types of firms, including specialist intermediaries and suppliers, leading to the creation and adaption of firms. Firms also go out of business and employees and resources move elsewhere [26, 111]. Lastly, firms innovate and spot new opportunities on the basis of their prior knowledge and ideas and new ideas they come across via their experience and through their social and business networks. Ideas become combined and recombined in new ways, leading to changes in activities, resource demands and relations.

The actions of firms in producing and consuming things have been modelled in terms of various input-output or production functions. Models of the mechanisms of specialization are not common, although economic theories of exchange, scale and scope efficiencies, and transaction-cost theory, point to the existence of various mechanisms related to specialization within and between firms that can be modelled in various ways. Some ABM of the evolution of specialists exist, including BankNet, in which intermediaries arise from among the initial actors on the basis of trading patterns, scale and scope efficiencies and transaction costs [93], and a dynamic model of role differentiation in social networks [33]. Experiments have also been carried out to simulate the evolution of trade and specialization that suggest ways of modelling and the various mechanisms involved [65]. Other general mechanisms such as learning, choosing and imitating have received far more modelling attention.

5.5.2 *Mating*

Potential trading partners encounter each other and choose, get chosen, accept or refuse to do business together and form relations. This could vary from random processes to ones that are influenced by past interactions, similarity, predispositions and communication networks. Changing or keeping partners depends on the evaluation, choice and learning processes of each party.

There are many examples of partner search and mating mechanisms in simulations, as it is one of the core mechanisms necessary to design a network simulation. Simple models of network generation use random pairing [37] or preferential attachment relative to the number of existing links [15]. Preferential attachment has been extended in many ways. Agents might lose attractiveness with age [27], or form relationships relative to activity [38] or performance [89] or based on prior successful cooperation [44]. Relationships can also be formed depending on expectations [100], or by copying links of established members [105]. Iterated games with choice and refusal of partners have been used to model the development of trading relations in which actors make and accept offers to trade on the basis of their previous experience with a partner [72, 97]. Attraction and gravity models mostly developed in geography are also relevant [4, 23], as is Schelling's (1971) classic model of the processes of segregation [94].

5.5.3 *Dancing*

Relations begin with initial interactions and exchange mechanisms. The experience and outcomes of business and social interactions and exchanges over time lead to learning and to changes in resources, attraction, perceptions, beliefs and evaluations for all those involved. Repeated interactions produce, reproduce and change a relationship atmosphere of some kind [113] that involves various types of actor bonds developing or not, for example, trust, commitment, power/dependence, cooperativeness, opportunism, empathy and understanding. These bonds affect willingness to engage in further interactions and exchange and the strategies used. Innovation also takes place in relations. Relationship termination can occur due to endogenous or exogenous changes, including interactions with third parties and the general environment.

Modelling the way firms interact over time has been dealt with mainly in terms of various types of learning models and iterated games. Learning has been modelled in a number of ways to reflect the ways people (and hence firms) adapt their perceptions of each other and how this affects their behaviour. These include changes in trust, power and cooperation [64, 102]. Axelrod and Hamilton's [9] early simulations focused on the iterated prisoner's dilemma and the conditions under which cooperative strategies emerge. Later work has considered other types of games and the evolution of strategies as a result of the experience of and performance in interactions over time [76, 100]. Similar models have been used to investigate the evolution of cooperation and the evolution of group and network structures under conditions of individual and group selection [56, 70, 85]. In a simulation of the Marseilles fish market, buying and selling agents learn to engage in and reward loyal behaviour. Sellers can reduce their risk caused by demand volatility and buyers benefit from better service and discounts [67].

5.5.4 *Interconnecting Relations*

Interactions and atmosphere in one relation can affect, positively or negatively, interactions in others that are connected to it [e.g., 5, 17, 108]. This is due to the operation of various kinds of mechanisms, including the way exchanges are interconnected positively or competitively along value chains [30–32]. In addition, communication and diffusion processes spread information and ideas through networks, leading to learning and adaption.

Many simulations depict the way in which relations are interconnected in terms of the effects of network structure on activities performed, including impacts on learning strategies and the profits of traders [69]. Cooperative behaviour in strategic games is facilitated if networks are sparsely connected [82]. Preferential attachment tends to connect hubs with each other and it has been found that cooperation can spread in this way to the entire network [92]. Network positions can affect individual payoffs [110] and the establishment of trust and reliability in one relationship can impede the success of another [64]. Networks for economic interactions and those for the exchange of information do not necessarily coincide. Diffusion models on networks also model the way information flows across networks [46]. Lastly, choice models necessarily involve linking relations when agents compare and evaluate alternatives partners.

5.5.5 *Environmental Impacts*

Business relations and networks operate in a context of a more general environment that both enables and constrains what happens. The initial conditions of any model are part of the historical environment, as they reflect the experience and outcomes of prior events that are not directly modelled. The environment can also be included in terms of patterns of change over time in key exogenous conditions, including the expansion and contraction of markets and resources, technological and infrastructure changes, and business laws and customs. This approach is exemplified in the modelling of the unfolding history of the Anasazi tribe over hundreds of years, in which the modellers included the known weather and topography as part of the environment [11]. Testing the implications of taxes and subsidies shows that they can facilitate the establishment of cooperative behaviour, which proves to be persistent, even after these are terminated [75]. More technical results show that the network structure is influenced by the payoffs of strategic games as well as the reliability of communication transmission [87]. Moreover, the relation between the time scales of actions on the network and the speed of the rewiring process of the network has been found to be important. A high rewiring speed of connections can essentially change the payoff structure of strategic games so that they favour cooperation [83]. In addition, the environment can be incorporated as its own model system that is coupled with the model of the focal business networks, as for example when a fish industry network is linked to an ecological model of fish stocks and fish behaviour, or to a broader economic model. The environment model responds to outcomes in the network model and generates events with which the network model has to cope.

5.5.6 *Discussion*

There are some areas in which additional modelling effort is required to represent particular mechanisms. Only a few examples could be found for the development of specialization in simulations. Although simulations can be initiated in terms of actors with different characteristics, roles and degrees of specialization, modelling the processes by which these differences emerge and change over time is not so well developed.

There are many examples of models of partner search and mating mechanisms because it is a core mechanism in network formation. Many simulations deal with search mechanisms and their impact on network topologies that mimic empirical processes. Nevertheless, there seem to be no simulations allowing agents to negotiate the terms of a relationship.

Business "dancing," the way firms interact, learn and adapt relations over time, has been dealt with in terms of various types of learning models and iterated games. Simulations do not yet capture the richness and multidimensional nature of the evolution of relationships. In most simulations, a connection either exists or not. Only [67] is an exception.

Various mechanisms exist that capture the ways the behaviour and outcomes in one relation are connected to that in other relations in a network, including general structural and network position effects, diffusion processes and choice processes that require conceptualization and inclusion rather than being confined to the types of connections identified in marketing theory (e.g., [5]).

Last, environmental impacts in terms of the heterogeneity or location of actors have not yet been included in simulations. Including such heterogeneity would allow a closer matching of simulations with actual business networks but necessarily restrict the generality of simulations. The environment can be controlled in various ways in terms of starting conditions, parameter values and imposed historical patterns of change in exogenous conditions, and can be the subject of its own model, which would then be coupled with the business network model.

5.6 Business Networks Agent-Based Modelling System

In order to design, develop, calibrate and evaluate ABM of business relations and networks, expert programming skills are required. This is so even though there have been major advances in the development of more user-friendly, general purpose ABM platforms, such as SWARM REPAST, MASON and NetLogo. All require knowledge and expertise in object-oriented programming languages such as Java, C++ or NetLogo. This represents a major barrier to research in business, economics and marketing, including business relations and networks, because academics and research students generally have little, if any, training in programming.

Business, economic and social researchers are trained for the most part in linear, variable-based comparative-static theory and methods rather than nonlinear, mechanism and process-based, dynamic and evolutionary theory and methods. The emphasis is on mathematics and statistics, which, although relevant to conceptualizing, designing, analysing and evaluating ABM, are not sufficient to build them. This situation may change, as future generations of academics and researchers grow up with computers and models of this type and more courses are offered to train them in ABM. Of course, researchers in business and economics can team up with expert programmers, but this limits their ability to become involved in the design and development of the ABM, and expert programmers have limited domain knowledge and insights to guide them.

To overcome this problem, we are engaged in developing an ABM system that can be used by researchers interested in studying business relations and networks that does not require programming expertise. We call this the Business Networks Agent-Based Modelling System (BNAS). This will provide researchers, as well as educators, managers and policy makers, with a user-friendly modular, menu-driven, modelling environment that will enable them to develop ABM relevant to their own interests, in which they can monitor, summarize and analyse the results to enable sharing and publication, and to contribute to the further development of BNAS. It will be similar in some ways to the design of modelling systems being developed to examine the behaviour of power markets [98]. BNAS will be embedded in a general ABM platform such as RePast or MASON. Researchers must understand the types of business relations and networks and theories they want to model and test, including the relevant mechanisms and rules governing the behaviour of the actors involved. The BNAS will be able to guide researchers in translating their theories and stylized descriptions of business relations and networks and their contexts into a functioning ABM in which they will be able to select among alternative ways of implementing various mechanism and rules for inclusion in the model, as well as being able to contribute new rules and mechanisms, and to be able to combine and link them into an integrated ABM model ready for analysis and testing.

The potential uses and value of BNAS include

- Building models to reproduce known structural, dynamic and evolutionary patterns, and to systematically examine their stability, behaviour and performance under counter-factual conditions that would be impossible to do in the field.
- Developing ABM to illustrate and test alternative theories of the dynamics and evolution of business relations and networks.
- Examining how different mechanisms and processes interact and perform under varying conditions.
- Examining and comparing results using alternative implementations of key mechanisms and features of actual or proposed business networks.
- Examining and testing the impact of different controllable (firm and policy) and noncontrollable factors on network dynamics, development and performance.
- Developing "flight simulators" for managers, industries and policy makers in order to sensitize them to the often counterintuitive outcomes of complex nonlinear systems and to explore the effects of different scenarios and types of environments, similar to uses in the military.

5.7 Conclusion

We have highlighted the nature, role and importance of business relations and networks to the functioning of business and the economy. We have also argued that research regarding the dynamics, formation and evolution of business relations and networks is underdeveloped yet essential if we are to develop scientifically sound advice and guidance for managers and policymakers. To make progress, we argue for the greater use of ABM methods to simulate CAS such as business relations and networks, and we show how this can be done in terms of identifying and modelling key causal mechanisms and processes.

We have reviewed the literature in business, economics and marketing to identify key mechanisms affecting the dynamics and evolution of business networks. These have been grouped into five broad types, and we have identified previous attempts to codify examples of these types of mechanisms. Although examples for each basic type are available, the congruence with empirically identified mechanisms is limited and requires further model development in order to represent them.

A barrier to the development of such ABM research is the limited training and expertise business, economics and social researchers have in relevant programming and ABM methods. To overcome this, we propose the development of a general purpose ABM system that does not require sophisticated programming expertise, yet allows researchers to become involved in the detailed design, construction, analysis and testing of their models. We believe a similar approach could be used to advance ABM in other domains because issues of dynamics, development, evolution and complexity are fast becoming a major area of attention in many research domains and in the world at large.

The growing relevance and interest in these issues is indicated most clearly with the award of the Nobel Prize in Economics to Elinor Ostrom in 2009 for her work on complex systems and the management of common pool resources, to Thomas Schelling in 2005 for his work on the unintended macro outcomes of micro behaviour in complex systems and to Vernon Smith in 2002 on the role of social relations and the self-organizing properties of market systems (not to mention the earlier award to Herbert Simon). The relevance and importance of the types of fine-grained nuanced models of complex systems that ABM can provide is further indicated by the failure of traditional economic models to predict the recent global financial crisis. Existing models make heroic assumptions about the nature and rationality of the behaviour of the representative actors in their models, which have been increasingly highlighted, leading to calls for more realistic models [e.g., 14, 51].

References

1. Abbott, A. (2001). *Time Matters: On Theory and Method* (University of Chicago Press, Chicago).
2. Adami, C. (1998). *Introduction to Artificial Life* (Springer, New York).
3. Aldrich, H. (1999). *Organizations Evolving* (Sage Publications, London).
4. Allen, P. M. and Sanglier, M. (1979). A dynamic model of growth in a central place system, *Geogr. Anal.* **11** (3), pp. 256–272.
5. Anderson, J. C., Håkansson, H., and Johanson, J. (1994). Dyadic business relationships within a business network context, *J. Marketing* **58** (4), pp. 1–15.
6. Ariño, A. and Torre, J. D. L. (1998). Learning from failure: towards an evolutionary model of collaborative ventures, *Organ. Sci.* **9** (3), pp. 306–325.
7. Augier, M. and March, J. G. (2004). *Models of a Man: Essays in Memory of Herbert A. Simon* (MIT Press, Cambridge, MA).
8. Axelrod, R. (2006). *Handbook on Research on Nature-Inspired Computing for Economics and Management*, ed. Rennard, J.-P., Chapter 7 "Advancing the Art of Simulation in the Social Sciences," (Idea Group, Hershey PA) pp. 90–100.
9. Axelrod, R. and Hamilton, W. D. (1981). The evolution of cooperation, *Science* **211** (4489), pp. 1390–1396.
10. Axtell, R. (2005). The complexity of exchange, *Econ. J.* **115** (504), pp. F193–F210.
11. Axtell, R., Epstein, J. M., Deand, J. S., Gumermane, G. J., Swedlundg, A. C., Harburgera, J., Chakravartya, S., Hammonda, R., Parker, J., and Parkera, M. (2002). Population growth and collapse in a multiagent model of

the Kayenta Anasazi in Long House Valley, *Proc. Natl. Acad. Sci. USA* **99** (supp. 3), pp. 7275–7279.

12. Axtell, R. L., Robert, A., Epstein, J. M., and Cohen, M. D. (1996). Aligning simulation models: a case study and results, *Comput. Math. Organ. Th.* **1** (2), pp. 123–141.
13. Balderston, F. E. and Hoggatt, A. C. (1962). *Simulation of Market Processes*. IBER Special Publications (Institute of Business and Economic Research, Berkeley).
14. Ball, P. (2010). Model citizens: building SimEarth, *New Sci.* 2784.
15. Barabási, A.-L. and Albert, R. (1999). Emergence of scaling in random networks, *Science* **286** (5439), pp. 509–512.
16. Bianconi, G. and Barabási, A. L. (2001). Competition and multiscaling in evolving networks, *Europhys. Lett.* **54** (4), pp. 436–442.
17. Blankenburg-Holm, D., Eriksson, K., and Johanson, J. (1996). Business networks and cooperation in international business relationships, *J. Int. Bus. Stud.* **27** (5), pp. 1033–1053.
18. Borgatti, S. P., Mehra, A., Brass, D. J., and Labianca, G. (2009). Network analysis in the social sciences, *Science* **323**, pp. 892–895.
19. Bowersox, D. J. (1972). *Dynamic Simulation of Physical Distribution Systems*. Msu Business Studies (Division of Research, Graduate School of Business Administration, Michigan State University, East Lansing).
20. Brenner, T. (2006). *Handbook of Computational Economics* **2**, eds. Tesfatsion, L. and Judd, K. L., Chapter 18 "Agent Learning Representation: Advice on Modelling Economic Learning," (Elsevier, Amsterdam) pp. 895–947.
21. Bunge, M. (1997). Mechanism and explanation, *Philosophy of the Social Sciences* **27** (4), pp. 410–465.
22. Buttriss, G. and Wilkinson, I. F. (2006). Using narrative sequence methods to advance international entrepreneurship theory, *J. Int. Entrepreneurship* **4**, pp. 157–174.
23. Clarke, M. and Wilson, A. G. (1983). The dynamics of urban spatial structure - progress and problems, *J. Reg. Sci.* **23** (1), pp. 1–18.
24. Dagnino, G. B., Levanti, G., and Destri, A. M. L. (2008). *Network Strategy* **25**, ed. Silverman, B., Chapter "Evolutionary Dynamics of Inter-Firm Networks: A Complex Systems Perspective," (Emerald Group Publishing Limited, Bingley, UK) pp. 67–129.
25. Di Paolo, E. A., Noble, J., and Bullock, S. (2000). Simulation Models as Opaque Thought Experiments, *Seventh International Conference on Artificial Life*. MIT Press, Cambridge, MA, pp. 497–506.

26. Dixon, D. F. and Wilkinson, I. F. (1986). Toward a theory of channel structure, *Res. Market.* **8**, pp. 27–70.
27. Dorogovtsev, S. and Mendes, J. (2000). Evolution of networks with aging of sites, *Phys. Rev. E* **62** (2), pp. 1842–1845.
28. Dwyer, F. R., Schurr, P., and Oh, S. (1987). Developing buyer-seller relationships, *J. Marketing* **51**, pp. 11–27.
29. Easley, D. and Kleinberg, J. (2010). *Networks, Crowds, and Markets: Reasoning About a Highly Connected World* (Cambridge University Press, Cambridge, UK).
30. Easton, G., Brooks, R. J., Georgieva, K., and Wilkinson, I. F. (2008). Understanding the dynamics of industrial networks using Kauffman Boolean networks, *Adv. Complex Sys.* **11** (1), pp. 139–164.
31. Easton, G., Wiley, J., and Wilkinson, I. F. (1999). Simulating Industrial Relationships with Evolutionary Models, *28th European Marketing Academy Annual Conference* (Humboldt University, Berlin).
32. Easton, G., Wilkinson, I. F., and Georgieva, C. (1997). *Relationships and Networks in International Markets*, eds. Gemünden, H. G., Ritter, T., and Walter, A., Chapter 17 "Towards Evolutionary Models of Industrial Networks - a Research Programme," (Elsevier, Oxford) pp. 273–293.
33. Eguíluz, V. M., Zimmermann, M. G., Cela-Conde, C. J., and Miguel, M. S. (2005). Cooperation and emergence of role differentiation in the dynamics of social networks, *Am. J. Sociol.* **110** (4), pp. 977–1008.
34. Ehrhardt, G. C. M. A., Marsili, M., and Vega-Redondo, F. (2006). Phenomenological models of socioeconomic network dynamics, *Phys. Rev. E* **74** (3), pp. 036106–11.
35. Epstein, J. M. (2006). *Generative Social Science: Studies in Agent-Based Computational Modeling* (Princeton University Press, Princeton).
36. Epstein, J. M. (2008). Why model?, *J. Art. Soc. Soc. Sim.* **11** (4), pp. 12.
37. Erdös, P. and Renyi, A. (1959). On random graphs, *Publ. Math. Debrecen* **6**, pp. 290–297.
38. Fan, Z. and Chen, G. (2004). Evolving networks driven by node dynamics, *Int. J. Mod. Phys. B* **18** (17-19), pp. 2540–2546.
39. Følgesvold, A. and Prenkert, F. (2009). Magic pelagic—an agent-based simulation of 20 years of emergent value accumulation in the North Atlantic herring exchange system, *Indus. Market. Manag.* **38** (5), pp. 529–540.
40. Ford, D. (1980). The development of buyer-seller relationships in industrial markets, *Eur. J. Market.* **15** (5/6), pp. 339–354.

41. Forrester, J. W. (1961). *Industrial Dynamics* (M.I.T. Press, Cambridge, MA).
42. Fronczak, P., Fronczak, A., and Holyst, J. A. (2006). Self-organized criticality and coevolution of network structure and dynamics, *Phys. Rev. E* **73** (4), pp. 046117–4.
43. Gavrilets, S., Duenez-Guzman, E. A., and Vose, M. D. (2008). Dynamics of alliance formation and the egalitarian revolution, *PLoS ONE* **3** (10), pp. e3293.
44. Gilbert, N., Pyka, A., and Ahrweiler, P. (2001). Innovation networks—a simulation approach, *J. Art. Soc. Soc. Sim.*, **4** (3).
45. Gilbert, N. and Troitzsch, K. G. (2005). *Simulation for the Social Scientist*, 2 Ed. (Open University Press, Berkshire, UK).
46. Goldenberg, J., Libai, B., and Muller, E. (2001). Talk of the network: a complex systems look at the underlying process of word-of-mouth, *Market. Lett.* **12** (3), pp. 211–223.
47. Gross, T. and Blasius, B. (2008). Adaptive coevolutionary networks: a review, *J. Roy. Soc. Interface* **5** (20), pp. 259–271.
48. Haase, M. and Kleinaltenkamp, M. (2011). Property rights design and market process: implications for market theory, marketing theory and S-D logic, *J. Macromarket.* **31**, pp. 148–159.
49. Håkansson, H., Harrison, D., and Waluszewski, A. eds. (2005). *Rethinking Marketing: Developing a New Understanding of Markets* (Wiley, San Francisco, CA).
50. Håkansson, H. and Johanson, J. (1993). *Industrial Networks* **5**, ed. Sharma, D. D., Chapter "Industrial Functions of Business Relationships," (JAI Press, New York) pp. 13–29.
51. Haldane, A. G. and May, R. M. (2011). Systemic risk in banking ecosystems, *Nature* **469** (7330), pp. 351–355.
52. Halinen, A. and Törnroos, J.-Å. (1998). The role of embeddedness in the evolution of business networks, *Scand. J. Manage.* **14** (3), pp. 187–205.
53. Hedström, P. (2005). *Dissecting the Social: On the Principles of Analytical Sociology* (Cambridge University Press, Cambridge, UK).
54. Held, F. (2010), Developing Agent-Based Models of Business Relations and Networks as Complex Adaptive Systems, *School of Marketing* (University of New South Wales, Sydney).
55. Held, F., Wilkinson, I. F., Marks, R., and Young, L. (2010). Exploring the Dynamics of Economic Networks: First Steps of a Research Project, *3rd World Congress on Social Simulation* (Kassel, Germany).

56. Henrich, J. (2004). Cultural group selection, coevolutionary processes and large-scale cooperation, *J. Econ. Behav. Organ.* **53** (1), pp. 3–35.
57. Hidalgo, C. A., Klinger, B., Barabasi, A.-L., and Hausmann, R. (2007). The product space conditions the development of nations, *Science* **317** (5837), pp. 482–487.
58. Holweg, M. and Bicheno, J. (2002). Supply chain simulation - a tool for education, enhancement and endeavour, *Int. J. Product. Econ.* **78** (2), pp. 163–175.
59. Huang, Y. (2010), Understanding Dynamics of Trust in Business Relationships, *School of Marketing* (University of New South Wales, Sydney).
60. Huang, Y. and Wilkinson, I. (2006). Understanding Trust and Power in Business Relationships: A Dynamic Perspective, *International Marketing and Purchasing Group Conference* (Milan, Italy).
61. Huff, D. L. (1964). Defining and estimating a trading area, *J. Marketing* pp. 373–378.
62. Iacobucci, D. and Hibbard, J. D. (1999). Toward an encompassing theory of business marketing relationships (BMRS) and interpersonal commercial relationships (ICRS): an empirical generalization, *J. Interact. Market.* **13** (3), pp. 13–33.
63. Jackson, M. O. (2008). *Social and Economic Networks* (Princeton University Press, Princeton, NJ).
64. Kim, W.-S. (2009). Effects of a trust mechanism on complex adaptive supply networks: an agent-based social simulation study, *J. Art. Soc. Soc. Sim.* **12** (3), p. 4.
65. Kimbrough, E. O., Smith, V. L., and Wilson, B. J. (2008). Historical property rights, sociality, and the emergence of impersonal exchange in long-distance trade, *Amer. Econ. Rev.* **98** (3), pp. 1009–1039.
66. Kinch, N. (1993). The Long-Term Development of Supplier-Buyer Relation, *International Marketing and Purchasing Group Conference* (Bath, UK).
67. Kirman, A. P. and Vriend, N. J. (2001). Evolving market structure: an ace model of price dispersion and loyalty, *J. Econ. Dynam. Control* **25** (3/4), pp. 459–502.
68. Kleijnen, J. P. C. (2008). *Design and Analysis of Simulation Experiments.* International Series in Operations Research and Management Science (Springer, Berlin).

69. Ladley, D. and Bullock, S. (2008). The strategic exploitation of limited information and opportunity in networked markets, *Computational Econ.* **32** (3), pp. 295–315.

70. Ladley, D., Wilkinson, I. F., and Young, L. C. (2007). Group Selection and the Evolution of Cooperation, *9th European Conference on Artificial Life* (Lisbon, Portugal).

71. Langton, C. G. ed. (1996). *Artificial Life: An Overview* Complex Adaptive Systems. (MIT Press, Boston).

72. Lebaron, B. and Tesfatsion, L. (2008). Modeling macroeconomies as open-ended dynamic systems of interacting agents, *Amer. Econ. Rev.* **98** (2), pp. 246–250.

73. Leombruni, R. and Richiardi, M. (2005). Why are economists sceptical about agent-based simulations?, *Physica A* **355** (1), pp. 103–109.

74. Li, S. X. and Rowley, T. J. (2002). Inertia and evaluation mechanisms in interorganizational partner selection: syndicate formation among U.S. investment banks, *Acad. Manage. J.* **45** (6), pp. 1104–1119.

75. Lugo, H. and Jiménez, R. (2006). Incentives to cooperate in network formation, *Computational Econ.* **28** (1), pp. 15–27.

76. Marks, R. E. (1989). Breeding Optimal Strategies: Optimal Behavior for Oligopolists, *Proceedings of the Third International Conference on Genetic Algorithms, George Mason University*. Morgan Kaufmann Publishers, pp. 198–207.

77. May, R. M., Sugihara, G., and Levin, S. A. (2008). Ecology for bankers, *Nature* **451**, pp. 893–895.

78. Mckelvey, B. (2004). Towards a complexity science of entrepreneurship, *J. Bus. Venturing* **19** (3), pp. 313–341.

79. Midgley, D. F., Marks, R. E., and Cooper, L. G. (1997). Breeding competitive strategies, *Manage. Sci.* **43** (3), pp. 257–275.

80. Mosekilde, E., Larsen, E. R., and Sterman, J. D. (1991). *Beyond Belief: Randomness, Prediction, and Explanation in Science*, eds. Casti, J. L. and Karlqvist, A., Chapter "Coping with Complexity: Deterministic Chaos in Human Decisionmaking Behavior," (CRC Press, Boston, MA) pp. 199–229.

81. Narayandas, D. and Rangan, V. K. (2004). Building and sustaining buyer-seller relationships in mature industrial markets, *J. Marketing* **68** (3), pp. 63–67.

82. Ohtsuki, H., Hauert, C., Lieberman, E., and Nowak, M. A. (2006). A simple rule for the evolution of cooperation on graphs and social networks, *Nature* **441**, pp. 502–505.

83. Pacheco, J. M., Traulsen, A., and Nowak, M. A. (2006). Coevolution of strategy and structure in complex networks with dynamical linking, *Phys. Rev. Lett.* **97** (25), pp. 258103–4.
84. Parke, A., Wasserman, S., and A., R. D. (2006). New frontiers in network theory development, *Acad. Manage. Rev.* **31** (3), pp. 560–568.
85. Pennisi, E. (2009). On the origin of cooperation, *Science* **325** (5945), pp. 1196–1199.
86. Poole, M., Scott, A. H., Van De Ven, K. D., and Holmes, M. E. (2000). *Organizational Change and Innovation Processes: Theory and Methods for Research* (Oxford University Press, New York).
87. Pujol, J. M., Flache, A., Delgado, J., and Sangüesa, R. (2005). How can social networks ever become complex? Modelling the emergence of complex networks from local social exchanges, *J. Art. Soc. Soc. Sim.* **8** (4), pp. 12.
88. Rauch, J. E. (2010). Does network theory connect to the rest of us? A review of Matthew O. Jackson's social and economic networks, *J. Econ. Lit.* **48** (4), pp. 980–986.
89. Ren, J., Wu, X., Wang, W. X., Chen, G., and Wang, B. H. (2006). Interplay between evolutionary game and network structure: the coevolution of social net, cooperation and wealth, arXiv:physics/0605250v2.
90. Ritter, T., Wilkinson, I. F., and Johnston, W. (2004). Firms' ability to manage in business networks: a review of concepts, *Indus. Market. Manag.* **33** (3), pp. 175–183.
91. Rond, M. D. and Bouchikhi, H. (2004). On the dialectics of strategic alliances, *Organ. Sci.* **15** (1), pp. 56–69.
92. Santos, F. C. and Pacheco, J. M. (2005). Scale-free networks provide a unifying framework for the emergence of cooperation, *Phys. Rev. Lett.* **95** (9), pp. 98104.
93. Sapienza, M. D. (2000). *Economic Simulation in Swarm*, eds. Luna, F. and Stefansson, B., Chapter 6 "An Experimental Approach to the Study of Banking Intermediation: The Banknet Simulator," (Kluwer, Boston) pp. 159–179.
94. Schelling, T. (1971). Dynamic models of segregation, *J. Math. Soc.* **1**, pp. 143–186.
95. Schweitzer, F., Fagiolo, G., Sornette, D., Vega-Redondo, F., Vespignani, A., and White, D. R. (2009). Economic networks: the new challenges, *Science* **325** (5939), pp. 422–425.
96. Simoni, M., Tatarynowicz, A., and Vagnani, G. (2006). The Complex Dynamics of Innovation Diffusion and Social Structure: A Simulation

Study, *WCSS 2006—The First World Congress on Social Simulation, Kyoto, Japan*.

97. Stanley, E., Ashlock, D., and Smucker, M. (1995). *Advances in Artificial Life*, 4.4 "Iterated Prisoner's Dilemma with Choice and Refusal of Partners: Evolutionary Results," (Springer, Berlin) pp. 490–502.
98. Sun, J. and Tesfatsion, L. (2007). Dynamic testing of wholesale power market designs: an open-source agent-based framework, *Computational Econ.* **30** (3), pp. 291–327.
99. Sydow, J., Schreyögg, G., and Koch, J. (2009). Organizational path dependence: opening the black box, *Acad. Manage. Rev.* **34** (4), pp. 689–709.
100. Tesfatsion, L. (1997). A trade network game with endogenous partner selection, *Comp. Approach. Econ. Prob.* pp. 249–269.
101. Tesfatsion, L. and Judd, K. L. eds. (2006). *Handbook of Computational Economics* 2. Handbooks in Economics Vol. 13. (Elsevier, Amsterdam).
102. Tomassini, M., Pestelacci, E., and Luthi, L. (2010). Mutual trust and cooperation in the evolutionary Hawks-Doves game, *BioSystems* **99** (1), pp. 50–59.
103. Van De Ven, A. H. and Engleman, R. M. (2004). Event- and outcome-driven explanations of entrepreneurship, *J. Bus. Venturing* **3**, pp. 343–358.
104. Van De Ven, A. H. and Poole, M. S. (2005). Alternative approaches for studying organizational change, *Organ. Stud.* **26** (9), pp. 1377–1404.
105. Vázquez, A. (2000). Knowing a network by walking on it: emergence of scaling, *arXiv:cond-mat/0006132v4*.
106. Watts, D. J. and Dodds, P. S. (2007). Influentials, networks, and public opinion formation, *J. Cons. Res.* **34** (4), pp. 441–458.
107. Welch, C. and Wilkinson, I. F. (2004). The political embededness of international business networks, *Int. Market. Rev.* **21** (2), pp. 216–231.
108. Wiley, J., Wilkinson, I. F., and Young, L. (2009). A comparison of European and Chinese supplier and customer functions and the impact of connected relations, *J. Bus. Ind. Market.* **24** (1), pp. 35–45.
109. Wilhite, A. (2001). Bilateral trade and small-world networks, *Computational Econ.* **18**, pp. 49–64.
110. Wilhite, A. (2006). *Handbook of Computational Economics* **2**, eds. Tesfatsion, L. and Judd, K. L., Chapter 20 "Economic Activity on Fixed Networks," (Elsevier, Amsterdam) pp. 1013–1045.

111. Wilkinson, I. F. (1990). Toward a theory of structural change and evolution in marketing channels, *J. Macromarket.* **10**, pp. 18–46.
112. Wilkinson, I. F. (2001). A history of network and channels thinking in marketing in the 20th century, *Australas. Market. J.* **9** (2), pp. 23–52.
113. Wilkinson, I. F. (2008). *Business Relating Business—Managing Organisational Relations and Networks* (Edward Elgar, Cheltenham, UK).
114. Wilkinson, I. F., Wiley, J., and Easton, G. (1999). Simulating Industrial Relationships with Evolutionary Models, *28th European Marketing Academy Annual Conference* (Berlin, Germany).
115. Woodside, A. G. ed. (2010). *Organizational Culture, Business-to-Business Relationships, and Interfirm Networks* 16. Advances in Business Marketing & Purchasing (Emerald Books, Bingley, UK).
116. Young, L. and Wilkinson, I. F. (2004). Evolution of Networks and Cognitive Balance, *International Marketing and Purchasing Group Conference* (Copenhagen, Denmark).
117. Zimmermann, M. G., Eguíluz, V. M., and San Miguel, M. (2004). Coevolution of dynamical states and interactions in dynamic networks, *Phys. Rev. E* **69** (6), pp. 065102.

Chapter 6

Agent-Based Modelling of Social Networks: Natural Resource Applications of Human/Landscape Interactions across Space and Time

Randy Gimblett,[a] Robert Itami,[b] and Aaron Poe[c]
[a]*School of Natural Resources and Environment, University of Arizona, Tucson, Arizona 85712, USA*
[b]*GeoDimensions Pty Ltd, 16 Tullyvallin Crescent, Sorrento, Victoria 3943, Australia*
[c]*Partnerships and Community Engagement, Chugach National Forest, 3301 C Street, Suite 300, Anchorage, Alaska 99503, USA*
gimblett@ag.arizona.edu, bob.itami@geodimension.com.au, apoe@fs.fed.us

6.1 Introduction

There is a growing body of theoretical and applied research focused within the context of human–environment interactions. What has been learned from these investigations is that conventional methods used in planning and management of human–landscape interactions fall far short of the needs of decision-makers who must evaluate the cascading impacts of humans in visitor landscapes. Many public land agencies, local governments and international organizations

Networks in Society: Links and Language
Edited by Robert Stocker and Terry Bossomaier

ISBN 978-981-4316-28-6 (Hardcover), 978-981-4364-82-9 (eBook)
www.panstanford.com

are exploring tools such as agent-based simulation models coupled with social science methods such as social networks that elicit responses from humans about their expectations, experiences, livelihoods, attitudes, preferences, family structure and so on and behavioural responses to resource management. There is a need to analyse and compare interactions and feedbacks between and within social and biophysical subsystems that serve to couple the human–natural subsystems in many types of landscapes. The goal is to identify commonalities in coupling mechanisms that govern the trajectory of human–nature linkages. A diverse array of modes of travel, experience opportunities, desired outcomes and benefits are evident in these landscapes. These outcomes vary in terms of the scale as well as the environmental and social context in which they are situated. Only by linking the social and environmental implications to the flow patterns generated by human pattern of use simulations is it possible to begin to properly manage the quality of experience for visitors in recreation settings. Although the application of simulation to study human–landscape interactions is in its infancy, there is need to develop a comprehensive and empirically based framework for linking the social, biophysical and geographic disciplines across space and time [6].

This issue is at the heart of a complex management problem explored in this paper in Prince William Sound (PWS) Alaska where outdoor recreation is an important part of the Alaskan lifestyle. Although many have chosen to live in Alaska to maintain such a lifestyle, out of state and international visitors who annually converge on the Alaskan landscape to hunt, fish and experience the wildness pose a threat to sustaining such a lifestyle. There is growing concern within the federal agencies mandated to oversee the management of this landscape in areas such as PWS that increased competition and rapid growth in sport hunting and fishing, both commercial and private as well as ongoing subsistence activities may be threatening the ability of the resource to sustain such use. Of equal concern is whether the very wilderness experiences that Alaskans and visitors are seeking are not equally being threatened. As recreational use levels increase in PWS (e.g., kayaking, wildlife viewing, pleasure boating, hunting, fishing, camping and so on), it is inevitable that encounter levels and associated impacts will increase, visitor conflicts will arise and native Alaskans could be displaced from traditional harvest areas. The Chugach National Forest (CNF) mandated to manage PWS is focusing its efforts to gain an understanding of the spatial and temporal patterns of recreation use in the Sound in order to better inform management objectives and ensure they are based on current and projected levels of use [14]. In order to compliment this information, the CNF also needs to consider a risk management approach to determine whether level of use exceeds quality of service standards for recreation, safety, environment, social and economic criteria that could focus on management alternatives that include use levels, safety, environmental, social and economic risks. This information will inform future use allocation strategies for visitors seeking to have limited contact with others.

Earlier consultancies in the PWS study, now referred to as the Sound, had established projected growth rates for commercial and recreational traffic; however, a defensible method for determining water or river-based capacity for the various forms of traffic had not been determined. The underlying assumption had been that some single metric such as "maximum number of vessels per mile" could be established to determine the overall capacity of the entire Sound. This definition does not recognize the very different physical operating characteristics of a variety of water-based users, or a basic understanding about the constituents surrounding and using the Sound as well as the quality of experiences required to complete wilderness solitude recreation. It is clear that a robust defensible way of defining capacity that takes into account water characteristics, user's values, vessel types and physical infrastructure had to be developed.

The purpose of this chapter is to explore the coupling of a social networking approach known as the level of sustainable activity (LSA) with agent-based modelling to explicitly evaluate the relationship between quality of experience and varying levels of density of user types from the perspective of local focus groups (e.g., paddlers, motorized vessels, hunters and fishers, and so on). This technique uses simulation technology to understand future patterns of human use and to make predictions about the intensity of use under a range of management scenarios on the basis of the information gained from this study.

6.2 Social Networks Informing Natural Resource Management

There has been much written about spatially explicit agent-based models over the last decade [5, 6, 8] and some discussion about coupling human and natural systems [6, 9, 11, 12] but little has focused on using any form of social networks and their linkage with these models. Social survey research and a handful of other methods have been used to capture data about interactions among and between social groups and their associated landscapes, and only a few studies have used these data to build spatially explicit models to study their interactions and predict future effects of these scenarios. Lauber et al. [10] point out that although there is a growing emphasis on effective natural resource management over the last 20 years in many landscapes, this has lead to significantly elevating attention on collaborative processes and increased reliance on collaboration. Social networks have only recently been used as ways to effectively engage and capture local knowledge about natural resource conflicts and issues surrounding resource management. For example, Duberstein [4] used a social network analysis approach to examine the social connectivity of small-scale fishing communities and the association of network structures with collaborative behaviour of these fisheries in the Northern Gulf of California. This study found considerable connectivity of communities via kinship ties of small-scale fishers, both within the region and to other areas in Mexico. The results suggest that an understanding of the social networks that connect communities and the potential pathways for information transfer, combined with a system of enforceable rules and policies and effective outreach methods and materials, may help managers and resource users more effectively and sustainably manage. Agent-based simulations could help in this instance to unravel the way the rules and policies could affect these interconnected communities using the Northern Gulf of California well into the future.

Although there is a rise in the inclusion and reliance on stakeholder groups to aid decision-making, there have been few formal frameworks or processes established to do so, let alone linked to formal models such as spatially explicit models that have been used to solve a plethora of resource management problems. The characteristics of these social networks (in which stakeholders and local constituents are involved, what roles they play and with whom they interact) may greatly influence their contributions to collaborative natural resource management. One example of a successful research project coupling social networks within a spatial framework is by Smith et al. [15] who developed a Bayesian Belief Network (BBN) model and linked it to a geographical information system (GIS) to map habitat suitability for the Julia Creek dunnart (Sminthopsis douglasi), an endangered ground-dwelling mammal of the Mitchell grasslands of north-west Queensland, Australia. Expert knowledge, supported by field data, was used to determine the probabilistic influence of grazing pressure, density of the invasive shrub prickly acacia (Acacia nilotica), land tenure, soil variability and seasonal variability on dunnart habitat suitability. The model was then applied in a GIS to map the likelihood of suitable dunnart habitat. Sensitivity analysis was performed to identify the influence of environmental conditions and management options on habitat suitability. The study provides an example of how knowledge of the intricacies of land tenure and associated operations, limited empirical data combined within a BBN model and linked to GIS data can assist in recovery planning of endangered fauna populations.

6.3 Dichotomies in Communication/Language between Disparate Interacting Groups

Dichotomies in communication between and within interacting groups exist in many if not every resource management situation. Conflict resolution is based on the fact that humans share environments and use those landscapes for a variety of reasons with varying perceptions, ethics and use levels. How humans communicate with each other, with resource managers and across a spectrum of desired activities and outcomes is the very foundation of much of the work currently undertaken in recreation management. A basic understanding of such a communication language and understanding is so fundamental that it is critical to deriving and implementing sustainable management plans [1]. Many authors such as Cole [2] have advocated for a more thorough understanding of recreation users, especially when it comes to understanding the nature of visitor densities and their effects on experience opportunities and how these opportunities might be compromised by interactions with other users. It has been shown by Cole [2] that a "better understanding of visitor experiences is also likely to create a richer vocabulary with which to articulate and differentiate between alternative experiences". Effective communication between recreation users and managers is fundamental to understanding and maintaining a quality spectrum of recreation experience opportunities.

The following cases present two examples of how an accurate understanding of the social networks that exist among communities of users in natural resource settings can lead to more effective decision making and management prescriptions.

In studying dispersed recreation commercial boating use along the Colorado river as it travels through Grand Canyon National Park, the importance of understanding the commercial social network and the rules that drive these interactions was discovered [7]. The many commercial boat operators dominate use along the river. In order to develop a simulation model of their travel patterns and behaviours for evaluating management strategies, a series of interviews were conducted with the hopes of deriving inherent rules from which these models could be developed. During the interview process, it was discovered that each company did not act alone on the river. In fact, the social network of these commercial operators was very complex and interdependent. Commercial operators in fact frequently met to cooperatively work out their trip schedules for the river to reduce interactions between groups and improve visitor experience. In addition, through the process of building this social network, it was discovered that there were very subtle interaction rules that had dramatic consequences on resource protection. For example, although the launch schedule could be used to control traffic and encounters on the river, commonly used by public land managers, the interactions and communication on the river was more important to developing a robust simulation that accurately portrayed river use. Boat leaders, as they encountered each other on the river, would communicate about the place each would camp for the night. As it ended up, boat leaders liked to camp together so that they could strategize and socialize, and as a result, parties converged on beaches or campsite locations that far exceeded their capacity, leading to adverse impacts on these sites. Here, the social network of commercial tour operators was the important driver for predicting patterns of human behaviour along the river. In discussions with the tour operators, it was also discovered that raft parties would send scouts down the river in advance of other parties to secure the sites that they liked to visit. These are excellent examples of how such subtle and intricate behaviour rules would never have been discovered without the understanding about how this complex social network of use was occurring. These rules contributed to a robust simulation that provided high predictive capability for future management of the river.

Learning the interdependencies between various recreation groups using PWS, Alaska is another good example of the importance of coupling human and environmental issues through social networks [14]. Inherent rules regarding this social network are prevalent. Commercial fishing with its large presence in various locations in the Sound tends to dominate the social interaction through their intensity of fishing and continual presence during peak fishing season. Other modes of travel avoid sprawling nets to minimize conflict. Hunters are in a continual conflict with kayakers who frequent the beaches that are patrolled from the water for black bear sightings. Direct conflict occurs when social groups overlap on the same set of beaches. One group is seeking a place to camp on shore for the night while the other is looking to hunt. Indirect conflict occurs between kayakers and hunters when they bear-bait the beaches to attract black bear into range. Kayakers indirectly happen upon these sites and can come into direct conflict with black bear who are ultimately attracted to the beaches. Understanding the social networks under which these recreation groups operate and in particular understanding the rules that govern their interaction is imperative to building an agent-based model to simulate their interactions and derive sustainable solutions for land managers.

6.4 The Level of Sustainable Activity Approach as a Formal Way to Capture Social Networks and Integrate into Agent-Based Models

In order to first understand the communication language within and between groups using natural resource settings and second to model these interactions over time to derive effective and sustainable solution, a common framework needs to be established that merges social and biological systems. A recent development in this area referred to as the LSA framework was developed to accomplish this task. This framework is a generalization of the level of service (LOS) concept developed by the Transportation Research Board. LOS is a measure by which transportation planners determine the quality of service for users with different travel modes as a means of defining travel network capacity. Roadway LOS is a stratification of travellers' perceptions of the quality of service in relationship to carriage width, traffic flows, traffic density and safety. Any given stretch of road can be managed for five LOS from A—very low density—allowing travellers to move freely at different speeds—to Level E—jammed where there are long delays, speeds are lowered and collision hazards are high. It is the goal of management through user research to determine the appropriate LOS to manage for. As each Level is quantified in terms of density, flow rates and quality of service, the framework can also be used to evaluate LOS for existing road networks. Although a motorist is, in general, interested in speed of his journey, LOS is a more holistic approach, taking into account several other factors. They can be safety, traffic density (or a measure of congestion), quality of experience and so on.

The concept of LOS for recreation might then be thought of as a management strategy for balancing recreation use with protection. The concept behind LSA is that capacity for an area is different for each user group and varies in relation to the physical setting, travel and user characteristics, the provision of facilities and the interaction with other users. LSA then can be thought of as a scale of end-user experience, a method for unravelling their social networks and communicating with other user groups to derive a common language from which to build sustainable management solutions. Each recreation setting has a range of service levels defined for each type of recreation activity from very low levels of use, with minimal environmental and social impacts to high-density use with high levels of user interaction, higher levels of potential environmental and social impacts, and more intensive facility and management requirements. The preferences of different recreation users are a key component of the management of recreation use and the definition of capacity. Each recreation user-type (such as rowers, commercial tour operators, water taxis and ships) has different requirements in terms of safety, ability to perform their intended activity and level of satisfaction on the basis of the mix and density of vessels sharing the river zone.

This framework provides a way to consider both the physical and social dimensions of recreation with clear targets for quality of service by user group. The LSA concept integrates physical characteristics of the setting, including environmental or social risks; defining recreation use densities for each LSA (A through E) for motorized and nonmotorized use; user preferences for levels of use for specific activities in specific settings, user attitudes toward competing user safety, environmental and social risk factors relating to increasing use densities and suggestions from users on management options for dealing with the above risks.

6.5 Level of Sustainable Activity Linked to Agent-Based Spatial Simulations

An understanding of the temporal and spatial distribution of visitor use is fundamental to many of the questions that planners and managers ask. Are the existing use patterns sustainable and appropriate to resource and experience goals? Are existing spatio-temporal distributions optimal for visitor experience, resource protection and efficient operations? Such questions cannot be answered without knowledge of the kinds, amount and distribution of visitor use. Particularly in large areas and in areas with complicated access and circulation patterns, planning staff may only have anecdotal information about use concentrations, lengths of stay in various areas, crowding, underused or overused facilities and other factors. Further, staff and public perceptions of use patterns are often at odds [3].

But essential in planning for management of public lands is that many of the impacts and tradeoffs associated with various planning alternatives are qualitative, value laden, and difficult to demonstrate. Yet, the public is asked to "buy in" to future conditions that can profoundly affect their visitation and experiences, sometimes without clear understanding of the implications of various choices. How might planning alternatives affect a visitor's ability to move about at his or her own pace? How might the alternatives affect visitors' chances of coming to the area whenever they want? What kinds of tradeoffs or sacrifices (using a reservation system or permit system, e.g., or riding a shuttle bus) might be asked of the public to protect resources? Are the goals "worth it" to the public? To answer these sorts of questions, more tools are needed to help understand and monitor baseline conditions, estimate appropriate use levels, describe the consequences of management alternatives and more effectively communicate these consequences to the public. Increasingly, visitor use simulation modelling is gaining recognition as a critically important tool for professional planning and management of recreation on public lands [3].

Recreation Behavior Simulator (RBSim) is one such tool. It is a software program fully integrated with ArcMap that has been specifically developed by researchers from GeoDimensions Pty Ltd and the University of Arizona for studying recreation problems. It has been used to solve many land management problems and in particular to model the distribution of recreation across diverse landscapes [5, 8].

When linked with the LSA framework, simulation provides precise management guidance on the relationship between visitor flow volumes, densities and LOS. Specifically, this type of framework provides detailed advice and informs management for maintaining and achieving quality of service objectives for quality of experience, safety, environmental protection, social interaction and enforcement. Simulations can inform management on existing peak use and forecast future use levels for specified time periods (e.g., 5 and 10 years). The outputs of the simulations provide a spatial view of the changes in visitor densities and volumes at peak use levels. The LSA results are the primary means by which the results of the simulations are analysed and evaluated. LSA results are used to interpret the results of the simulations to determine if and when a management zone reaches capacity in terms of users perceptions of safety and experience. These simulations demonstrate how well the proposed quality standards will work, given increases in diversity and visitor numbers for an area. To demonstrate the merging of the two techniques, the following case study is presented.

6.6 Prince William Sound Case Study

PWS is located in south-central Alaska at 61° N, 148° W. The Chugach and Kenai Mountain ranges separate most of PWS from interior Alaska and two large islands, Montague and Hinchinbrook, shelter the hundreds of bays and islands that make up PWS from the Gulf of Alaska [13]. The maritime climate of PWS is characterized by heavy annual precipitation, much of which falls in the form of snow during long winters. Summers are generally cool and wet. The 7000 km of convoluted shoreline are composed of tall, rock cliffs, gravel beaches, tidal flats, rocky outcrops and islands, estuaries and tidewater glaciers. PWS shorelines are exposed to large fluctuations in tide (+6 to −1 m) and different levels of wave action.

A project was undertaken to provide a contemporary analysis of user experience in PWS in order to (a) evaluate existing management situation to determine whether users are experiencing the qualities/attributes for which managers have planned; (b) assess the recovery of the recreation/tourism service, which was impacted by the spill and currently listed as "not fully recovered" by the Trustee Council; and (c) describe human use patterns and understand the potential for displacement resulting from competition between user groups (i.e., injured services, such as subsistence and recreation tourism) as well as evaluate the intensity of overlap with resources injured by the spill. The project evaluates user activities, impacts, conflicts and experiences (including concerns and perceptions) in the Sound.

The study involved the distribution of $n = 1377$ questionnaires to recreation users departing from Whittier, Valdez and Cordova during the summer of 2008. This survey posed questions to users about their motivations, desired opportunities, satisfaction with recreation experience relative to their expectations, encounter dynamics and asked respondents to log their trip using a map document diary. In addition, reported use locations from a 2005 trip diary study targeting recreation users in the Sound were evaluated. Questionnaire and trip diary reported use locations were compared with empirically identified use locations from transect and shore-based observations.

Previous efforts dependent on the use of simulation software (Recreation Behavior Simulator, RBSim) to define patterns of use by dispersed recreationists were successful with a sample of nearly 5–12% of all users entering a region through a limited number of access points [6]. In other studies in landscapes with remote access and low numbers of visitors, it was found that a return rate of 30% is typical [14]. Although this does not usually meet social science standards for sampling, this is sometimes the reality of working in these environments. An agent-based simulation was developed from the trip diaries and survey responses to evaluate the distribution of use across the Sound by mode of travel, analyse popular destinations, evaluate how long visitors spend at these sites and who they encounter both on water and on land. A series of simulations of existing peak use and forecast future use levels in the Sound into the future were undertaken. The simulation was replicated 100 times to acquire a more accurate representation of distribution of visitor use. For each simulation replication, the following simulation variables are randomized:

- Selection of trip itineraries—a trip is randomly drawn from a pool of trips with the same port and the same day of week.
- Launch times are randomized.
- Travel speeds are randomized around a mean for each vessel type.
- Departure times from destinations are randomized within the time frame of 1 h.

Simulation outputs were then averaged over the 100 replications. As many trips are multiple day trips, it was necessary to run the simulation a sufficient time in order to fully populate the simulation before outputs are calculated. In this simulation, the first 2 weeks of June are used as the warm period and are therefore discarded before simulation outputs are calculated on the remaining simulation days.

The outputs of the simulations provide a spatial view of the changes in visitor densities and volumes at peak use levels. The LSA framework (described below) is the primary means by which the results of the simulations are analysed, evaluated and interpreted. This step was critical in understanding the social network of user and their interdependencies in the Sound. LSA results are used to determine if and when the management zone reaches capacity in terms of users perceptions of safety and experience. Questions regarding varying levels of use density were posed to focus groups composed of hunters, recreational power boaters and kayakers using the LSA framework of inquiry. Focus groups were organized for inquiry sessions in the communities of Cordova, Valdez and Anchorage. They were presented with representative human use scenarios for three different subregions within the Sound known to have varying levels of existing use on the basis of prior studies. They were asked to provide perspectives on the use levels presented as they related to their ideal, expected and maximum tolerable scenarios of use for three different types of users.

6.7 Limits of Sustainable Activity Focus Groups to Understand Social Network of Recreation Users

In this study, the LSA framework was implemented to explicitly evaluate the relationship between quality of experience and varying levels of density of user types from the perspective of local focus groups in this case study. LSA was developed to:

- Define quality of experience objectives and criteria for competing user groups in outdoor recreation settings.
- Understand social interactions and networks within and between user groups.

- Using this knowledge to define visitor capacity levels within and between user groups.
- Elicit management suggestions from users for attaining or maintaining quality of experience objectives.

LSA workshops were conducted in three communities around the Sound:

- Cordova is a small remote community with a population of 2242 (2008) accessible only by boat (including a ferry service) or plane.
- Valdez is a small remote community with a population of 3787 (2008); it is accessible by boat, road and plane.
- Anchorage is Alaska's largest city with a population of 279,243 (2008); it is well serviced by all transportation modes and is the source of most boat traffic into PWS.

Within these three communities, three recreational user groups were selected with separate workshops for each group for a total of nine workshops:

- Kayakers ($n = 23$)
- Recreational motor and sail boaters ($n = 22$)
- Hunters ($n = 7$)

In total, $n = 62$ people participated in the workshops.

An important aspect of managing quality of experience in outdoor environments is determining the relationship between use levels and quality of experience. A standard measure of use levels is density or the number of people, or in this case vessels or parties per square kilometer. In the LSA approach, five levels are designated from Level A (no use) to Level E (high use levels). These levels must be defined in the context of the level of traffic and environment under study. For example, it would be inappropriate to use the same density levels for an urbanized bay as for PWS, as the essential experience in PWS is wilderness. To determine realistic LSA levels for the Sound, user survey data from 2005 were used [16] in conjunction with the 2008 study described above. Daily use levels for kayaks, small motorized boats (<30 feet) and large motorized yachts and sail boats were examined for peak summer days for

Blackstone Bay, Unakwik Inlet and Sheep and Simpson Bays. Table 6.1 shows the density of vessels for each of the three analysis areas expressed in terms of the number of vessels per analysis area for each LSA level and the equivalent density in vessels per square kilometer. Note that the density for each LSA level is the same across all three analysis areas. Also, note that the levels from B to E are in a geometric progression with density doubling with each level. Once these density levels were determined, images were prepared for each analysis area using survey locations observed over the entire season in 2005 and 2008, and the appropriate number of vessels was randomly selected for LSA Level E for each analysis area. Levels D, C and B were then generated by randomly deleting vessels according to the numbers shown in Table 6.1.

Participants in each user group were asked to make evaluations in three different contexts:

Table 6.1 LSA density levels for three analysis areas

LSA level	Number of vessels	Vessels/sq km
Sheep and Simpson Bays	*68.478 sq km*	
A	0	0.00
B	2.1	0.03
C	4.2	0.06
D	8.5	0.12
E	17	0.25
Unakwik Inlet	*109.94 sq km*	
A	0	0.00
B	3.4	0.03
C	6.8	0.06
D	13.6	0.12
E	27.3	0.25
Blackstone Bay	*64.16 sq km*	
A	0	0.00
B	2	0.03
C	4	0.06
D	8	0.12
E	16	0.25

- Preferred or *Ideal LSA*
- *Expected LSA* at busy times
- *Maximum Tolerable LSA*

After making these evaluations, each user group was asked the following questions:

- What are the impacts of other (types) of users on your experience?
- What suggestions do you have for management to help you attain your desired experience?
- Are their issues such as safety, noise, environmental impacts or annoying behaviour of other users that we have not discussed?
- Other issues/questions

The LSA is simply a method for understanding the social networks that exist among the user groups and as a means for them to communicate their goals, preferences, desires and outcomes on resource management issues. Results are used to interpret the results of the simulations outlined above to determine if and when the management zone reaches capacity in terms of users perceptions of safety and experience. These simulations will demonstrate how well the proposed quality standards will work, given increases in diversity and numbers of users across the Sound. Being able to forecast future human use distributions in PWS will enhance the longevity of management decisions made under the upcoming capacity analysis.

6.8 Results from Applying the Level of Sustainable Use Workshops for Evaluating Management Scenarios

The following sections are only a sampling of the results found in this study to demonstrate the connection between social networking (workshops) and agent-based simulation.

For *Expected Peak Season LSA*, the evaluations were more variable. The reason behind this variability is that some users had not visited some of the locations, so opted out of the evaluation or made an evaluation on the basis of "what they imaged" the traffic to be. The users who had visited the area may or may not have experienced the traffic during peak periods. Generally, however, the highest densities were assigned to Blackstone Bay attributed to the close proximity to Whittier (and therefore with the easiest access to the greatest population) and Sheep and Simpson Bays because these bays are smaller with relatively high use levels than the area of the bay. Unakwik Inlet had the lowest ratings primarily due to the large size of the inlet and the relatively low use.

The median values for *Maximum Tolerable LSA* ratings were fairly consistent between communities for each bay. Also, the evaluations were fairly consistent by vessel type. Anchorage Kayakers have the highest tolerance for other kayakers in Blackstone Bay (LSA D). This is largely due to the large size of the bay and the fact that this level of use is already expected for Blackstone Bay. Valdez Kayakers also had high tolerance for other Kayakers in Sheep and Simpson Bays. Anchorage Kayakers had the lowest tolerance for other vessel types at Unakwik Inlet (LSA b for both small motorized boats and Yachts and Sailboats). Cordova Kayakers also had low tolerance for Yachts and Sailboats at Unakwik Inlet. Otherwise, there is a high level of consistency for other vessel types by Kayakers in all three communities with an LSA rating of c for small motorized boats and motorized yachts and sailboats for all three locations.

When one compares the Expected LSA for each vessel type and each location with the *Maximum Tolerable LSA* for the corresponding locations, it is evident that *Anchorage Kayakers* perceive that use levels across all locations for all vessel types are currently at capacity. In other words, they would not tolerate heavier use during busy times. If confronted with these levels, they would either be displaced to quieter locations or they would try to schedule trips at alternative times. Some Anchorage Kayakers indicated that there are already locations they avoid because of the unacceptably high use levels. The evaluations for *Ideal LSA* were consistent across all communities for all vessel types with high value placed on low density use (LSA a – no other vessels or b. 0.03 vessels/sq km) across all three locations. These ratings are consistent with the high-value Kayakers expressed for solitude and wilderness experience in the Quality of Experience discussion.

The picture for Valdez and Cordova Kayakers in comparing Expected LSA to *Maximum Tolerable LSA* is more complex. Blackstone Bay is at or over capacity for all vessel types. Sheep and Simpson Bays are at capacity for Motor yachts and sailboats and at capacity (Expected LSA = C, *Maximum Tolerable LSA* = C) or at capacity for small motorized boats for Valdez Kayakers or 1 LSA level under capacity for Cordova Kayakers (Expected LSA = c, *Maximum Tolerable LSA* = d). At Unakwik Inlet, Valdez and Cordova Kayakers perceive that there is excess capacity for small motorized boats during peak periods (Expected LSA = b, Maximum Tolerable = c). Cordova Kayakers feel that Unakwik Inlet is at capacity during peak periods for motorized yachts and sailboats (LSA = b), whereas Valdez Kayakers LSA would tolerate LSA Level c compared with *Expected Peak Season LSA* of b.

6.9 Level of Sustainable Activity: Recreational Motor Boats, Yacht and Sail Boats

The USFS managers had the greatest variation in LSA scores for *Ideal LSA* and *Maximum Tolerable LSA* reflecting the diversity of interests in this group. It was apparent during the workshop that some USFS managers are subsistence hunters and fisherman and their scores are more similar to those of the hunter groups discussed in the next section. Other USFS managers were more representative of the recreational boaters, though all place high value on solitude.

Generally, recreational boaters have a more social orientation than either Kayakers or Hunters; this is reflected in the generally higher LSA ratings for *Ideal LSA* than the other user groups. The reasons expressed for the higher ratings include the concept that other boats in the area are good (in case of emergencies) and that Kayaks cause little or no conflict. Valdez boaters had higher *Ideal LSA* ratings for Kayaks and small motorized boats at Blackstone Bay. Their reasoning was that this is the busiest bay and they expect it to be busy and it may be appropriate to have heavy traffic in this setting. The perception that kayaks have little impact on the feeling of wilderness is also reflected in the high LSA ratings for kayaks by all three recreational boating communities for allocations. Motor

Yachts and Sailboats are seen to have more impact than motor boats of the same size; however, recreational boaters as a group are much more tolerant of other boats than Kayakers and Hunters across all vessel types and all locations.

Maximum Tolerable LSA ratings are generally high for Blackstone Bay for Kayaks, as recreational boaters do not see them as having a big impact either on traffic or on sense of solitude. Also, there is an expectation that Blackstone Bay will be busy because of the proximity to Whittier and the opportunity to view glaciers. Respondents tend to be more tolerant of higher use levels in areas where higher use is expected and are known destinations for commercial tour boats and recreational boats. One of the key factors determining LSA levels for large boats is the availability of safe anchorages. Local knowledge and maps and guidebooks that show anchorages are valuable resources for recreational boaters in PWS. No values are reported for Valdez hunters, as invited members of the community failed to attend the workshop. For Anchorage and Cordova hunters, there is a high level of agreement on the *Ideal LSA* with level a and b ratings across all locations for all vessel types. This is consistent with the expressed value hunters place on solitude and their preference for low competition when hunting. Anchorage hunters expressed concern that PWS was getting too much pressure from hunters and that they were already getting displaced by heavy hunting use or depletion of game stocks.

6.10 Level of Sustainable Activity: Comparisons between Groups

When one compares and contrasts the three communities and the three recreational user groups, there are some distinct similarities and differences. Generally, Hunters have the highest requirements for solitude and the lowest tolerance for competition. Generally, they prefer "1 boat per bay" and this "unwritten law" is generally acknowledged and respected among subsistence hunters in local communities such as Cordova. However, conflict with other hunters occurs when this concept is not recognized, especially by commercial hunting guides. The key issues relating to the hunter groups have to do with the management of the hunting resource. Anchorage hunters felt that wildlife stocks were overhunted and that the practice of bear-baiting was undesirable. Hunting pressures, primarily from sports hunters as opposed to subsistence hunters, are considered to be intolerable during the peak hunting season, which causes local hunters to either move from traditional hunting areas or hunt during the shoulder seasons. Hunters are generally in PWS in the spring and fall and therefore do not overlap on the main summer recreational boating season. They therefore see little conflicts with Kayaks and other recreational boats so the main competition is with other hunters.

Kayakers follow hunters in the high value they place on solitude. This group values "low impact" camping, self-sufficiency, camaraderie with other kayakers and contact with nature. They are generally tolerant of other Kayakers because of the quiet mode of transportation and strongly shared values of low-impact recreation. However, they are less tolerant of small motorized boats because of the noise, speed, wake and the impact it has on quiet and solitude, and if they camp onshore, they compete for campsites. They are more tolerant of the larger motor yachts and sailboats because these boats are self-contained and do not compete for camping sites. Valdez kayakers were particularly opposed to the practice of bear-baiting because of the visual impact of litter from baked goods and potential danger posed by bears attracted to shoreline campsites.

Recreational boaters attending the LSA workshops were generally sailing motor yachts and sailboats. Smaller motorized boats such as cabin cruisers and runabouts and trailer boats were not represented among the respondents. Although this group, like the others, value solitude and wilderness experience as the "Ideal" LSA, they are, as a group, more socially oriented. They see little conflict or competition with Kayaks and have higher tolerance to small motorized boats than hunters or Kayakers. They have equal or higher tolerance to Kayakers with regard to Motor Yachts and Sailing boats. For this group, the availability of anchorages is critical. Because they are self-sufficient and not reliant on onshore campsites, they tend not to have conflicts with Kayakers, and when they travel, Kayakers tend to hug the shoreline, whereas recreational boaters and especially motorized yachts and sailboats use the open water. This group is quite aware of safety issues and mentioned the problem of the dangers for inexperienced sailors in any sized motor boat running out of fuel, unaware of where to find safe shelter in inclement weather, poorly equipped or maintained boats, and inadequate seamanship given the remoteness of many of the destinations in PWS.

When comparing all three communities and user groups, it is apparent that all groups value feelings of solitude. With exception of USFS managers, the median rating for *Ideal LSA* across all bays, for all communities for all recreational groups was a (no boats) or b (0.03 vessels/sq km). Those assigning a rating of a liked the idea of "being the first one there" or the feeling of "having the place to yourself". Those assigning a value of b felt that a few other vessels were positive for safety reasons. Valdez recreation boaters rated Blackstone Bay LSA c because they felt that Blackstone Bay was unique as a tourist destination because of its proximity to Anchorage and because of the opportunity to view glaciers and wildlife in an already busy bay.

In most cases, users' median ratings for *Expected Peak Season LSA* were higher, or much higher than *Ideal LSA*. This is expected given that *Ideal LSA* ratings were either a or b. However, *Maximum Tolerable LSA* ratings varied considerably between recreation types. Kayakers mostly assigned LSA c for *Maximum Tolerable LSA* across all bays and all vessel types with a few b and d ratings. Hunters also had relatively low *Maximum Tolerable LSA* ratings with mostly LSA b and c ratings for all locations. Recreational boaters had higher rating across all bays and vessel types with median *Maximum Tolerable LSA* ratings of c and with and even a few e ratings. These ratings are consistent with what users expressed in the workshops for Quality of Experience criteria with Recreational boaters being more social than kayakers or hunters.

6.11 Summary

Provided above is an example of how to couple human and natural systems using a form of social networks and agent-based models. There is much that still needs to be learned on how to acquire human knowledge through the use of social networks and how to build rules from these networks that can feed an agent-based model capable of mimicking local and institutional actions. This paper has demonstrated:

- Without a statistically representative sample of the users of the Sound, the social network of users and knowledge surrounding their preferences, patterns of use and desired conditions would not have been known leading to an unsustainable management plan.
- As the communication between groups using the Sound and management was nonexistent, it would have been nearly impossible to understand the issues surrounding patterns of use and ideal, preferred and tolerable density of use without the LSA framework.
- Combining the knowledge gathered through the surveys into an agent-based simulation framework allowed the team to fully understand the areas of concern and intricate relationships between groups and seasonal use patterns of those using the Sound.
- The LSA results that were subsequently used to interpret the outputs of the simulations allowed the team to determine if and when the management zone reaches capacity in terms of users perceptions of safety and experience. These simulations also demonstrate how well the proposed quality standards will work, given increases in diversity and numbers of users across the Sound into the future.
- The LSA workshops brought out similarities and differences within each of the networks of recreation groups. These differences were revealed, and through the LSA process, a consensus was derived regarding ideal, appropriate and maximum tolerable densities of use between the three groups using the sampled areas.
- Being able to forecast future human use distributions in PWS through the application of this framework will enhance the longevity of management.

6.12 Conclusion and Controversial Research Issues Identified and Future Research Directions

It is clear from our research that there is a demonstrated need to develop a comprehensive and empirically based framework for linking the social, biophysical and geographic disciplines across space and time. When coupled with the power of agent-based simulations, the LSA framework which is founded in social network analysis can provide precise management guidance by location, user group and time. As clear objectives of environmental quality, risk, perceived quality and so on are established at the onset, capacity for any recreational setting can be determined because it includes both the physical and social dimensions of recreation with clear targets for quality of service by user group. Rather than relying only on experts, which has been problematic in the past, LSA embraces the power of those who use the environment to aid in determining appropriate levels of use and simulation to evaluate alternatives and examine future scenarios. Most importantly, the LSA framework linked to agent-based simulations is very adaptable to any visitor landscape and any type of use. In addition, this combination of techniques can provide valued information for establishing appropriate capacities in visitor landscapes: balancing high-quality recreation experience with long-term protection of the landscape.

Our future research in the Sound will address the need to develop more refined sets of rules for our agent simulations through a more thorough examination of written transcripts of interviews, conversations and stories by extracting key linguistic cues that might indicate strategic/tactical thought processing. These can be done for each of the stakeholders in the Sound. The written and verbal statements will be examined to extract further perspectives on conservation, sustainability, livelihood, and recreational pursuits. A project is currently in progress referred to as 'Sound Stories' endeavours to collect this type of data, in the words, images, voices and science of the people who live, work in and love the Sound (http://www.explorethesound.org/). Local communities with strong ties to their natural setting and resources are being asked to share their stories. These are stories about heritage, community, events, experiences, observations and descriptions of the close ties these communities have to the land. These stories are collected through a dynamic and interactive website that also serves to educate a broad and diverse audience about the region, conservation the Sound's resources and guide people to enjoy the Sound in a sustainable way in the face of increasing use of the Sound. The overriding hope of the Sound Stories is to form a network of people and organizations to help protect the Sound's resources and assist Sound users to play a role in the conservation of the Sound. From these stories, there is a hope of extracting such perspectives that might not normally be captured in traditional survey and interview methodologies from which a more detailed linguistic analysis could be undertaken. Revealing these perspectives has potential to better inform and refine the decision-making inputs and evaluate resulting strategies from our agent simulations. Much more research and developmental work needs to be undertaken to explore methods to reveal these linguistic patterns.

In conjunction with this linguistic analysis, we will also explore the use of conditional probability BBNs as a method for encoding rules that govern social networks. When integrated within the ABM, they can enhance the heuristic and intelligent decision-making capabilities. BBNs could provide the capabilities to evaluate, derive and enhance system learning and, ultimately, to simulate future scenarios that result from rules governing local social and governing institutional conditions. BBNs have successfully been used to represent spatial decision-making processes, but have seldom been incorporated directly into a simulation environment such as an ABM or used to encode linguistic cues as probabilistic rules that might lead to more refined decision making. A truly coupled human–natural system is a seamless integration between rules governing land-use decisions, the models representing landscape change and intelligent decision-making systems capable of mimicking local and institutional actions. BBNs could provide a conceptual understanding of coupled system relations and calculate joint probabilities for decision options and outcomes of management policies.

References

1. Aplin, K., Kamal Azad, A., Al Bachchu, A., Baker, A., Belmain, S., Harun, M., Hasanuzzaman, A., Islam, M., Quraishi Kamal, N., Meyer, A., Mian, Y., Mohammod, N. and Singleton, G. (2004). *Ecologically-based rodent system for diversified rice-based cropping systems in Bangladesh—Evaluation of PETRRA-funded part of study*. Natural Resources Institute, University of Greenwich, Chatham, Kent, UK. 39pp.
2. Cole, D. N. (2001). Visitor Use Density and Wilderness Experiences: A Historical Review of Research. In Freimund, W., Cole, D. N. comps 2001. *Visitor Use Density and Wilderness Experience: proceedings*. (2000). June 1–3, Missoula, MT. Proceedings RMRS-P-20. Ogden, UT. U. S. Department of Agriculture, Forest Service, Rocky Mountain Research Station. 67pp.
3. Cole, D. N., Cahill, K., Hof, M. (2005) Why Model Recreation Use? In: Cole, David, N. (compiler). *Computer Simulation Modeling of Recreation Use: Current Status, Case Studies, and Future Directions*. Gen. Tech. Rep. RMRS-GTR-143. Ogden, UT: U.S. Department of Agriculture, Forest Service, Rocky Mountain Research Station. September 2005. 75pp.
4. Duberstein, J. N. (2009). The Shape of the Commons: Social Networks and the Conservation of Small-scale Fisheries in the Northern Gulf of California, Mexico. Unpublished Ph.D. Dissertation. School of Natural Resources and Environment. University of Arizona, Tucson, Arizona.
5. Gimblett, H. R. (2002). *Integrating Geographic Information Systems and Agent-Based Modeling Techniques for Simulating Social and Ecological Processes*. Oxford University Press, London.
6. Gimblett, H. R. and Skov-Petersen H. (eds). (2008). *Monitoring, Simulation and Management of Visitor Landscapes*. University of Arizona Press. 2008. 457pp.
7. Gimblett, H. R. (2005). Simulation of Recreation Use Along the Colorado River in Grand Canyon National Park. pp. 27–30. In: Cole, David N. (compiler). *Computer Simulation Modeling of Recreation Use: Current Status, Case Studies, and Future Directions. Gen. Technical Report RMRS-GTR-143*. Ogden, UT: U.S. Department of Agriculture, Forest Service, Rocky Mountain Research Station. September 2005. 75pp.
8. Itami, R., Raulings, R., MacLaren, G., Hirst, K., Gimblett, H. R., Zanon, D. and Chladek, P. (2003). RBSim 2: simulating the complex interactions between human movement and the outdoor recreation environment. *J. Nat. Conserv.* **11**, pp. 278–286.

9. Lassoie, J. and Sherman, R. (2010). Promoting a coupled human and natural systems approach to addressing conservation in complex mountainous landscapes of Central Asia. *Front. Earth Sci. China* **4**(1) pp. 67–82.
10. Lauber, B., Decker, D. and Knuth, B. (2008). Social networks and community-based natural resource management. *Environmental Management* **42**, pp. 677–687.
11. Liu, J., Dietz, T., Carpenter, S., Alberti, M., Folke, C. Moran, E., Pell, A. N., Deadman, T. Kratz, J. Lubchenco, E. Ostrom, Z. Ouyang, W. Provencher, C. L. Redman, P., Schneider, S. H. and Taylor, W. (2007). Complexity of coupled human and natural system. *Science* **317**(5844), pp. 1513–1516.
12. Monticino, M., Acevedo, M., Callicott, B., Cogdill, T., Ji, M. and Lindquist, C. (2004). Coupled Human and Natural Systems: A Multi-Agent Based Approach. Complexity and Integrated Resources Management. In Pahl-Wostl, C., Schmidt, S., Rizzoli, A. E. and Jakeman, A. J. (eds), *Transactions of the 2nd Biennial Meeting of the International Environmental Modelling and Software Society*, Manno, Switzerland, ISBN 88-900787-1-5.
13. Murphy, K. A., Suring, L. H. and Iliff, A. (2004). *Western Prince William Sound human use and wildlife distribution model, Exxon Valdez Oil Spill Restoration Project Final Report (Restoration Project 99339)*, USDA Forest Service, Chugach National Forest, Anchorage, Alaska.
14. Poe, A., Gimblett, H. R. and Itami, R. M. (2010). *Evaluating the Recreation Service Recovery: Evaluation of Prince William Sound User Experience, Exxon Valdez Oil Spill Restoration Project Final Report.* USDA Forest Service, Chugach National Forest, Anchorage, Alaska.
15. Smith, C., Howes, A., Price, B. and McAlpine, C. (2007). Using a Bayesian belief network to predict suitable habitat of an endangered mammal – The Julia Creek dunnart (Sminthopsis douglasi). *Biological Conservation* **139**(3-4) pp. 333–347.
16. Wolfe, P., Gimblett, H. R., Itami, R. and Garber-Yonts, B. (2008). Monitoring and Simulating Recreation Use in Prince William Sound, Alaska. In Gimblett, H. R. and H. Skov-Petersen (eds) (2008). *Monitoring, Simulation and Management of Visitor Landscapes.* University of Arizona Press. pp. 349–369.

Chapter 7

Social Media Networks and the "Unthinkable Present": A Users' Perspective

John Carroll[a] and David Cameron[b,*]
[a] *Centre for Research in Complex Systems, Charles Sturt University, Bathurst, New South Wales 2795, Australia*
[b] *Centre for Teaching and Learning, University of Newcastle, Newcastle, New South Wales 2300, Australia*
jcarroll@csu.edu.au, david.cameron@newcastle.edu.au

7.1 Introduction

A decade ago the Canadian author William Gibson observed that science fiction is often mistakenly credited with predicting the future, simply because technological change seems to happen so quickly. With the benefit of hindsight, he argues, observations of emerging trends can only seem prescient if they are not interrogated too deeply: "As I've said many times before the future is already here, it's just not very evenly distributed" [1]. What we perceive

*Supported by Teaching Fellowship with Charles Sturt University's Flexible Learning Institute.

Networks in Society: Links and Language
Edited by Robert Stocker and Terry Bossomaier

ISBN 978-981-4316-28-6 (Hardcover), 978-981-4364-82-9 (eBook)
www.panstanford.com

as new technology is often a combination or application of current but hitherto distributed knowledge or tools—for example, the relatively rapid development of smartphones and tablet computers can be attributed to many decades of prior development in telecommunications, computing and even photography and satellite navigation. What we have seen in the first decade of the 21st century is a coming together of existing social and computing networks to form new patterns of connections in the online world. These principles of human social interaction, painstakingly unearthed in the past by social scientists using small sample sizes and in-depth field research, are now becoming available for empirical research in an unprecedented way.

Gibson, who coined the term *cyberspace* in his novel *Neuromancer* [2], felt that a networked world would generate new patterns of social engagement. However, rather than trying to predict the future he now often mines an "unthinkable present" in his works of fiction. As a successful author he makes a living using science fiction as the tool to snap together the scattered pieces of a contemporary reality that many of us find difficult to observe or imagine as we go about our lives. As much as we might wish it to be so, daydreaming and soothsaying are probably not actually specified in most of our job descriptions.

In 1995 *Time* magazine devoted an entire issue to exploring the concept of cyberspace,[a] which Gibson describes in *Neuromancer* as

> a consensual hallucination experienced daily by billions of legitimate operators, in every nation, by children being taught mathematical concepts ... A graphic representation of data abstracted from banks of every computer in the human system. Unthinkable complexity. Lines of light ranged in the nonspace of the mind, clusters and constellations of data. Like city lights, receding ... [2].

More pragmatically, cyberspace can be defined as "a worldwide network of computer networks that use [protocols] to facilitate data transmission and exchange" [3]. The publication of the cyberspace

---

[a]The cover and contents of this March 1, 1995, edition of *Time* can be viewed at http://www.time.com/time/magazine/0,9263,7601950301,00.html

issue of *Time* magazine was significant in that it was the first time that mainstream media had embraced the notion and what it represented. The idea that "the information superhighway is a two-way street" was touted by a writer in the issue, David S. Jackson, who observed that media outlets and organizations were beginning to regularly encounter, and pay close attention to, the views of their readers or customers [4].

Like Gibson, the author Cory Doctorow notes science fiction's tendency to "predict the present," through what he calls a sense of "radical presentism":

> Mary Shelley wasn't worried about reanimated corpses stalking Europe, but by casting a technological innovation in the starring role of Frankenstein, she was able to tap into present-day fears about technology overpowering its masters and the hubris of the inventor. Orwell didn't worry about a future dominated by the view-screens from 1984, he worried about a present in which technology was changing the balance of power, creating opportunities for the state to enforce its power over individuals at ever-more-granular levels [5].

Perhaps "radical presentism" in the "unthinkable present" is overstating the case for this apparent forecasting. But even serious scientific types can attempt to predict the future based on current technology with some accuracy. Almost a decade before Gibson was outlining his approach, the U.S. inventor Ray Kurzweil was describing his vision of the Age of Intelligent Machines [6], an Industrial Revolution 2.0 that would be characterized by computer-based machines to extend and multiply our mental abilities.[a]

This second industrial revolution, he said, would "ultimately have a far greater impact than the [first industrial] revolution that merely expanded the reach of our bodies" [6]. It would be characterized by computer-based machines that would extend and multiply our mental abilities through the projection of our networked personas,

---

[a] Marc Prensky, whose term "digital native" was popularized to describe claimed cognitive shifts in younger learners, now refers to a broader, cross-generational approach to developing digital wisdom: "referring both to wisdom arising from the use of digital technology to access cognitive power beyond our innate capacity and to wisdom in the prudent use of technology to enhance our capabilities" [7].

exactly the process of online identity that was already beginning to happen. Kurzweil's ideas followed Intel co-founder Gordon E. Moore's projection in 1965 that computer processing power would double each year, and would continue to do so indefinitely [8].

While Kurzweil's predictions might read like science fiction, he too cannily bases his forecasting on observations of past and present trends, ranging from the evolution of life on earth to what became known as Moore's law. Evolution and advances in computing power are both, Kurzweil argues, aided by an ability to build on what already exists in the world rather than having to start from scratch.

The exponential nature of change expressed by Moore's law has clear implications for any network or activity that becomes dependent on these computing technologies. In line with both Moore and Kurzweil's predictions, many aspects of modern life in modern industrialized societies are now heavily influenced by networked digital technology. There has been a paradigm shift in the production and consumption of information [9], marked by the emergence of a digital, networked environment characterized by technological fluidity and frequent innovation, and driven by the exponentially increasing processing power of computers.

The field of social media networks is bounded by constant movement and debate in both theory and practice, spurred by constant technological developments and their cultural and social consequences [10, 11]. It is this shift in the availability of networked data that is making the invisible world of human social networks now available for inspection through the traces left by individuals within social media sites (Kleinberg, 2008). An emphasis can now be placed on how people use online social networks and cultural technologies, and the possibilities this provides for "human thinking, feeling and communication in a new medium" [3].

Before any specific networking characteristics are described, two features of this emerging digital networked culture are becoming evident: first, remediation, which is the remixing of old and new media [12], and second, bricolage, which is the assembly of available disparate media forms and exists online "in terms of the highly personalised, continuous and more or less autonomous assembly, disassembly, and reassembly of mediated reality" [9]. According to Deuze,

> the manifold scrambled, manipulated, and converged ways in which we produce and consume information worldwide are gradually changing the way people interact and give meaning to their lives. The emergence of a fragmented, edited, yet connected and networked worldview in itself is part of digital culture [9].

As the production and consumption of information become ever more reliant on digitization it is clear that individuals, as both producers and consumers of information, can be thought to be active in the process of meaning-making, engaging in remediation and becoming "bricoleurs" [9]. This twin process of remediation and bricolage along with the proliferation of media platforms and applications make any definitive views of network relationships within social media very difficult to codify into diagrammatic form. One way to see a current snapshot of how social utility and social interaction operates is to consider how individuals can use these tools in a functional way. Figure 7.1 is a diagram by Roger Harris [13] that attempts to define the functions social media users have available to them. It is an ecology based on a ladder of social media activities proposed by Li [14] that categorizes users as creators, critics, collectors, joiners, spectators and inactives.

7.2 Mapping Social Media: Verbs, Nouns and Collective Nouns

The range of interactions individuals engage in within social media applications is metaphorically presented in Fig. 7.1 as a wiring diagram, and it connects the ways people use social media to the types of social media sites available and then through to typical examples of the particular application that users are familiar with. This diagram represents an attempt to connect a functional ecology (verbs) of behaviours at the level of interaction rather than concentrate on the sometimes fleeting examples (nouns) of the individual applications that are usually shown in social media user based diagrams. So in network terms it demonstrates how social media verbal functions such as "create" are linked to specific platform such as blogs, social networks, forum groups, wikis and

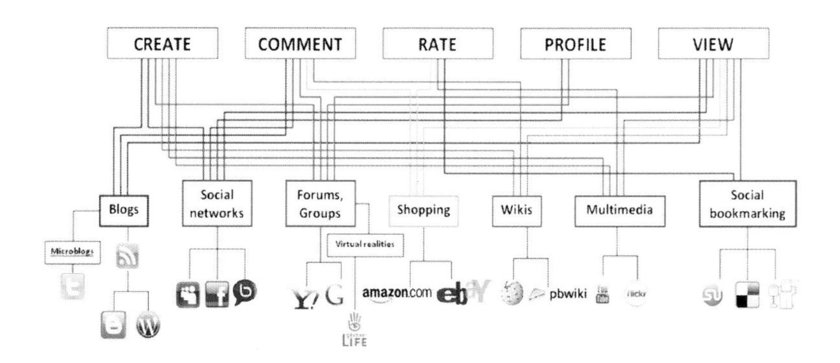

Figure 7.1 Social media ecosystem mapped as a wiring diagram [13].

multimedia rather than just directly to individual application forms like Twitter or Facebook. This allows a multiple view of how users utilise a range of platforms within a social media ecosystem.

The diagram in Fig. 7.1 is designed to demonstrate how the functions (we call them verbs) and the platforms (the nouns) interconnect through generic functional platforms types (or collective nouns, to continue our analogy). For example,

> We can see at a glance that shopping platforms allow comments and rating, and also views. They do not offer user generated content or social profiles. Likewise, social bookmarking offers views and ratings, but does not allow upload of UGC (user generated content), or social profiles [13].

The generic functional types (collective nouns) allow for multiple cross-platform use as Harris demonstrates, while the structural connections and the limitations of the connectivity of each platform is clarified within the overall social media ecosystem. The following sections explain how each functional platform (collective noun) operates in detail.

7.2.1 *Blogs and Microblogs*

If we now examine the first collective noun level of the diagram in Fig. 7.1, it shows a crucial generic platform that has developed in cyberspace in the last decade. This is the weblog or, as it has become universally known, the blog. This is essentially a web page with the latest post at the top, where the author publishes links and commentary around those posts [3, 15, 16]. When writing a blog, an individual

> simply logs on to their weblog account and uses a form to add text, hyperlinks, images, audio files and other objects. The software then typically formats the material in HTML, records a date at the top of the entry, and inserts the text at the top of the user's weblog, pushing previous entries down the page. . . . [A weblog] is also characteristically rich in hypertext links to other sites, and indeed the term "web log" and the statements by early "bloggers" suggest that this kind of webpage developed as a record of the user's latest browsing, which was made available for others' interest [17].

Most blogs pull together their content from a diverse array of other sites, in a method that can be likened to pastiche or bricolage [18], resituating information and social and cultural interaction within a larger context [16]. The advent of the blog is directly related to the development of the World Wide Web, and relies on an important characteristic: that of the "hyperlink," links that "may take users further in depth into an item, or to related stories or websites outside the news site for further information" [17].

Non-institutional blogging offers an alternative narrative of the citizen [16]. Blogs can be thought of as an embodiment of the contradictions of postmodernity. They may marginalize interest groups and cater to partisan ideals, but they may also encourage the creation of virtual networks and communities in which people can freely voice their opinions [16]. Some cultural and literary theorists see blogs as personal narratives [16] that offer alternative ethnographic perspectives, a new form of commentary, even an entire genre in themselves, that disrupts traditional narratives of history and could reinvent national identity and mediated memory [19]. Ibrahim explains further:

> New media to a degree resists the dominant power structures of mainstream media thus encouraging civic participation and

> plurality in the new spaces where audiences can appropriate the role of producers; where the pronoun "I" can assume a counterpoint to the dominant discourses in society [19].

In 1999, a number of public blogging tools were released, causing a rapid growth in the "blogosphere" or blogging community. One widespread application is Blogger, which allows individuals to create their own blog space very rapidly and with very low entry barriers to participation in the blogosphere. In scouring the Web and arranging topical links on one page, bloggers provided a valuable resource in navigating the huge amount of information available on the Web. The word *blog* entered the *Oxford English Dictionary* in 2003, and in that same year, a furore erupted over the status of blogging. Blogging critics say that most bloggers do not sub-edit, fact-check, or make any attempt at objectivity [20]. Supporters of blogging maintain that standards are slipping in the mainstream media and that news organizations need to embrace the Web and the new forms of dialogue it offers to users and be more transparent and accessible.

According to Boler, the ability of digital social and cultural networks to transmit the vast array of alternative and dissenting political commentary such sites as Blogger provide can, perhaps, be seen as a sign of "new counter-public spaces that coincide with increased mainstream media control and erosion of civil liberties surrounding free speech" [21]. Others say that the worldwide repository of millions of different viewpoints on the Internet makes winning the propaganda war significantly more difficult. According to Owen Gibson, "propaganda in the historical sense is simply not an option . . . when you can see opposing views at the click of a mouse" [22].

More recently, microblogging has emerged as a rapidly developing form as Twitter exploded in popularity. Often misrepresented as simply as a space to create, comment or view content, its ability for individuals to follow others makes it a highly efficient specialized content distribution medium through the agency of highly skilled microbloggers who disseminate up to the minute data in specialist fields through URL postings.

While there is also little in the literature about microblogging site Twitter (the popular online global social microblogging network with millions of users and continuing growth), interest is also increasing in academic areas about the impact and use of the network. According to Huberman, Romero and Wu, "many people, including scholars, advertisers and political activists, see online social networks as an opportunity to study the propagation of ideas, the formation of social bonds and viral marketing, among others" [23]. In terms of the growth and development of online news and digital media, Twitter is interesting in that many users of the service use it to report the latest news, or comment about current events [24], and it is being used by many organizations and individuals in this way also.

The impact of Twitter as a social network is also particularly significant given its rapid growth in terms of individual users of the service. On March 14, 2011, an official Twitter blog reported that an average of 460,000 new accounts had been created per day during the previous month.[a] Katie Stanton, vice president of international strategy at Twitter, announced in April 2011 that the service had passed its 200 millionth account.[b] Some analysts were quick to point out that it is difficult to judge how many Twitter accounts are actively used at any point in time.[c] Kwak, Lee, Park and Moon, in a study presented at the 19th International World Wide Web Conference, analyzed the entire Twitter site "and obtained 41.7 million user profiles, 1.47 billion social relations, 4,262 trending topics, and 106 million tweets" [25]. In their analysis, which they claimed was the first quantitative study on the entire Twittersphere and the information diffusion within it, they concluded that of all the tweets registered, more than 85% were news-related in some way, and therefore the service resembled not a social network so much

---

[a] http://blog.twitter.com/2011/03/numbers.html

[b] Stanton is quoted in live coverage of the Guardian Activate 2011 conference at http://www.guardian.co.uk/media/pda/2011/apr/28/guardian-activate-2011-new-york

[c] Typically, media and analysts seeks ways to assess the number of currently active Twitter users, versus the total number of accounts created, as in this March 2011 example from Business Insider: http://www.businessinsider.com/chart-of-the-day-how-many-users-does-twitter-really-have-2011-3

as a news network [25]. The ongoing effect this will have on existing news network structures is unclear.

Research suggests that the blurring of public/private boundaries on social media networks extends to blogs as, according to Viégas: "There is a disconnect between the way users say they feel about the privacy settings of their blogs and how they react once they experience unanticipated consequences from a breach of privacy" [26]. Viégas suggests that there should be clearly articulated social norms for blogging; more sophisticated controls over access to blog entries and more prominent indicators of the presence and number of readers on blog sites so that authors are reminded of their audience.

7.2.2 *Social Networks and Participatory Culture*

At the level of social networks within Fig. 7.1, the public can now create, comment, profile and view material as members of digital social and media networks as noted above both through their own personal blogs (e.g., Blogger or Blogspot) but also through a newer narrative form: participatory journalism. Participatory journalism generally falls into the broad categories of audience participation at mainstream news outlets such as the *New York Times* on the Web. There are also collaborative and contributory media sites such as Slashdot, Kuro5hin and Metafilter, as well as independent news and information websites such as the Drudge Report.

This growing trend comes at a time when traditional news media companies are coming under fire for the perceived concentration, conglomeration and globalization of ownership; with profits being maximised at the expense of news values. There is a direct link between this and the forms of "alternative" information networks that exist in the online landscape and the emerging "bottom-up" information initiatives such as IndyMedia and WikiNews.

Indymedia, or the Independent Media Centre, reports on social and political issues at a grassroots level, and was established as a reporting network to cover the demonstrations against the World Trade Organisation (WTO) meeting in Seattle in 1999. According to Meikle, "Its key feature was offering news coverage supplied by anyone who wanted to contribute, using free software and ideas from the Australian activists who had created the Active network" [27]. The software, developed by the Catalyst tech collective in Sydney, was created initially to service an online hub for Sydney activists that was launched in January 1999 and became a prototype for Indymedia: part calendar, part meeting place, part street paper [27].

Indymedia's open publishing model poses a unique set of problems, including libelous allegations and accusations of opinionated content, but its supporters maintain that the model is valid and should be upheld. The site's managers do not set the agenda—reporters and readers themselves decide what should be covered—but emphasize reading its contents with a critical eye [20].

7.2.3 *Forums, Groups and Convergence*

The word *convergence* in relation to groups and forums has several meanings, and is an important term when assessing the paradigm shift occurring in digital social and cultural networks. It can refer to multiple content platforms, the integration of previously distinct information industries or the migratory behaviour of information audiences [10]. It concerns the way consumers interact with the ever-increasing range of content available in the digital sphere and how this access to information is causing a significant cultural shift.

It also refers to the massive growth in social interaction, and sites such as Google Groups or Yahoo Groups provide ample space for the construction of personal identity through the information available in the digital sphere. As Henry Jenkins says, people use the application of this information "in the daily quest for understanding of the world and ourselves" [10].

One of the first theorists to discuss the word *convergence* in an academic context is Ithiel de Sola Pool, who has been called "'the prophet of media convergence" [10]. Pool wrote, "A process called 'convergence of modes' is blurring the lines between media, even point-to-point communications, such as the post, telephone and telegraph, and mass communications, such as the press, radio and television" [28]. As a component or concept applicable to the networked digital information landscape itself, convergence encompasses the transitional period we are currently experiencing, marked by unpredictability and a lack of clear direction. To borrow a colloquialism, this is a period of the "mashup," or the hybridization of content, where the elements of the online landscape are being mixed and matched in ways never seen before, and audiences are reforming and converging in new ways.

Much contemporary discourse in this area starts and ends with what Jenkins calls the "Black Box Fallacy": the notion that networked consumers of the future will access all their information through a single unit either in their homes or in a mobile device [3, 10]. Instead, Jenkins says, black boxes are proliferating, and this reflects the incompatibility between digital technologies, in that the pull towards more specialized devices exists at the same time as the push towards more generic media devices:

> We can see the proliferation of black boxes as symptomatic of a moment of convergence: because no one is sure what kinds of functions should be combined, we are forced to buy a range of specialised and incompatible appliances. On the other end of the spectrum, we may also be forced to deal with an escalation of functions within the same media appliance, functions that decrease the ability of that appliance to serve its original function, and so I can't get a [mobile] phone that is just a phone [10].

Convergence changes the dynamics that operate between audiences, information genres and industries, within marketplaces, and between existing technologies [10]. It is a process: the process of how consumers consume their information, and how industries respond to this consumption. Consumers are more aware of the forces behind their information consumption.

Convergence could be an opportunity for conglomeration or it could fragment markets as consumers, for example, move from television to the Internet. Major corporations are attempting to reinvent themselves in this climate of uncertainty, and consumers are undergoing a radical behavioural shift, a shift that "may have implications for how we learn, work, participate in the political process, and connect with other people around the world" [10].

The literature reflects the lack of an appropriate discourse for this converging and the wider impact it is having, and will continue to have. Some theorists borrow conceptual frameworks from other disciplines in their attempt to appropriate an applicable discursive model. "Media" is the plural of "medium," so its meaning could be interpreted in the sense that a medium is a carrier of something. Alternatively, a biologist would say a medium is a mixture of nutrients used to grow tissue cultures, or living organisms [11]. In this sense, Naughton says, you could treat human society or culture as

> an organism which depends on a media environment for the nutrients it needs to survive and develop. Any change in the environment—in the media which support social and cultural life—will have corresponding effects on the organism. Some things will wither; others may grow; new, mutant organisms may appear. The key point of the analogy is simple: change the medium, and you change the organism [11].

The concept of convergence can therefore be extended through the concept of an online networked "ecology," explored earlier in Fig. 7.1. As Naughton says,

> Organisms prey on one another; compete for food and other nutrients; have parasitic or symbiotic relationships; wax and wane; prosper and decline. And an ecosystem is never static. The system may be in equilibrium at any given moment, but the balance is precarious. The slightest perturbation may disturb it, resulting in a new set of interactions and movement to another—temporary—point of equilibrium [11].

This increasing domination of networked social media means that the old "push" model of information networks, where producers create and control content and push it to passive recipients, is being replaced by a "pull" medium, which involves dynamic, active information consumption and increased transparency for corporations—the emergence of "a truly sovereign, informed consumer" who actively produces as well as consumes information [11].

This trend highlights the possibility that there will come a point where users will create more content than media companies. It is possible that in the future, culture will be produced mostly by individuals with computers, software and an Internet connection, rather than large companies who can afford the high capital costs of production, as was the case only a decade ago. Corporations in future will need to address the challenges of trying to find business models that function within this new environment:

> One doesn't have to be a Marxist to realise that such a radical shift in the means of production will, in due course, impact on what Marx called the "superstructure"—the culture that sits atop the fundamental economic realities of production [11].

7.2.4 *Public Privacy and Shopping*

Uncertainties also exist around the status of citizen privacy in this age of participatory interaction through networked technologies especially when it is related to online purchases. Barber expresses concern about how new technologies, seen on the one hand as tools of liberation, could also conversely be used as tools of repression:

> As consumers tell shopping networks what they want to buy and tell banks how to dispense their cash and tell pollsters what they think about abortion, those receiving the information gain access to an extensive computer bank of knowledge about the private habits, attitudes and behaviours of consumers and citizens. This information may in turn be used to reshape those habits and attitudes in ways that favour producers and sellers working the marketplace or the political arena [29].

The notion of the Internet as "panopticon," in reference to the philosopher and social theorist Jeremy Bentham's model for efficient prison and neighbourhood control [30] can be extended into digital networks. In a panopticon, the observer can watch the prisoners without them being conscious of being watched: some commentators see the monitoring of Internet users through various means as proof that the Internet is similar to a panopticon. As Brignall explains,

> The panopticon as conceptual structure can be applied to any physical structure that provides the ability of those in a position

> of authority to monitor the "inmates" without the "inmates" knowing when they are being monitored. What is unique within the structure of the Internet is that it allows multiple layers of observation to occur such that the "inmates" can become the observers of other inmates. In such a situation, no one knows who is the observer and who is the observed [31].

In illustration of this notion, Tseng and Eischen discuss a US government proposal to create a Total Information Awareness surveillance system for all national digital communications [32]. They explain that while such a system may be largely technically unworkable, there was a mistaken general belief that the architecture of the Internet itself prevented such a system, as there were no physical access points through which all the data on the system flowed, using the United States as an example:

> In reality, north America does have central access points—six to be exact—through which all data moves because it is physically impossible to create redundant systems. This simple factor of geography potentially shapes policies on speech, privacy, terrorism, and government-business relations to name just a few [32].

The boundaries of social media spaces on the Internet, in terms of what information is public and what is private, are unclear and the illusion of privacy can often create boundary problems [33].

The concern over the privacy of individuals using the Internet extends to the wildly proliferating social networking platforms of recent times: Google Buzz, Bebo, MySpace, Facebook, Twitter and so on. Facebook, in particular, has been the subject of much speculation in terms of how it affects individuals' privacy, with research showing that some users have no awareness of how to control the information they disseminate on the network or how to control the visibility of their network profile [34, 35].

Criticism of Facebook has mounted due to the regular iterative changes the company has made to its privacy policy, changes that many see as an abuse of trust in terms of its users. According to Opsahl,

> Facebook originally earned its core base of users by offering them simple and powerful controls over their personal information. As Facebook grew larger and became more important, it could have chosen to maintain or improve those controls. Instead, it's slowly but surely helped itself—and its advertising and business partners—to more and more of its users' information, while limiting its users' options to control their own information [36].

There is clearly mounting interest in Facebook and social networking software in general among academics [37]. And, with its many millions of users, "we cannot deny that Facebook has a political impact—or at least a potential impact" [37].

7.2.5 *Wikis*

In 1994 the computer programmer Ward Cunningham set about creating a fast way to collaboratively publish Web content. He named his new application a wiki, drawing on the Hawaiian verb "wikiwiki," meaning quick or in a hurry. Effectively, the basic principles of any wiki are to

- allow all users to edit any page or create a new page within the website, using a standard Web browser;
- promote meaningful topic associations by simple and effective page-linking processes; and
- to involve the visitor in a collaborative and dynamic Web publishing landscape [38].

The most successful "noun" or example of these principles at work is the wiki-based online encyclopedia, Wikipedia. These principles are both the strength of Wikipedia as an open and user-created resource, and the main source of criticisms leveled at it about the veracity of its content and its utility as a scholarly repository of shared human knowledge. Another term associated with such open source software is "wabi sabi" [39], which refers to a Japanese world view or aesthetic system, centered on the acceptance of transience—a very apt association when considering wiki applications. The Wikipedia citation is used here knowingly, in this sense, as an illustration of the wabi sabi principle.

Although functionally a wiki is an easy form of content creation and sharing, the collaborative and social side built around the group development of a shared resource lends itself to the development of communities of shared interest. Wikis exist around almost any imaginable theme or topic, though arguably few are as well known or regularly accessed as Wikipedia. It can be seen that while functionally it is an online encyclopedia, it is also the product of an idealized social activity with an underlying universal aim:

> Wikipedia is not merely an online multilingual encyclopedia; although the Web site is useful, popular and permits nearly anyone to contribute, the site is only the most visible artifact of an active community... Most importantly, it sometimes reveals what I call a good faith collaborative culture. Wikipedia is a realization—even if flawed—of the historic pursuit of a universal encyclopedia: a technology-inspired vision seeking to wed increased access to information with greater human accord [40].

Tapscott and Williams [41] use the term "wikinomics" to describe the impact of collaboration and participation on organizations, with wikis just one manifestation of the Web as an "internetworked constellation of disruptive technologies" that produces a scramble to find responses to emerging phenomena at an increasing rate:

> Previous technology-driven revolutions, like the electrification of industry, took the better part of a century to unfold. Today the escalating scope and scale of the resources applied to innovation means that change will unfold more quickly. Though we are still just beginning a profound economic and institutional adjustment, incumbents should not expect a grace period. The old, hardwired "plan and push" mentality is rapidly giving way to a new, dynamic engage and co-create economy. A hypercompetitive global economy is reshaping enterprises, and political and legal shifts loom [41].

The intrinsic openness of the basic wiki design, with a purpose-built infrastructure that accommodates and encourages collaborative global development of peer-reviewed content, illustrates the potential for these new social media forms to radically change traditional producer/audience relationships.

7.2.6 *Multimedia*

Similarly there are online tools where people can share photos and video clips, create material and comment on and swap favourite Web links, and share lists of their favourite books or films. The central functions that drive multimedia networks are the ability of users to create content, rate others' content and view an enormous range of similar material.

One of the key features of these social media tools is the ability for both producers and consumers to describe this range of content using natural language keywords:

> These systems enable users to add keywords (i.e. "tags") to Internet resources (e.g., web pages, images, videos) without relying on a controlled vocabulary. Tagging systems have the potential to improve search, spam detection, reputation systems, and personal organization while introducing new modalities of social communication and opportunities for data mining [42].

As the barrier to entry in creating, rating or viewing material is so low, this provides a positive incentive to participation within a wide range of online communities of interest. Clay Shirkey has recently extensively researched this phenomenon in his text *Cognitive Surplus* (2010).

7.2.7 *Social Bookmarking*

In the context of Web-based media, the process of "bookmarking" emerged as a means of creating, editing and storing a personal list of websites and links within the browser software. As users began to rely on this ability to collect and organize their favourite online resources, the functionality of this feature increased to include, for example, an ability to export and import a bookmark list, or to describe entries with keywords to enhance searching through long lists. A limitation of this browser-based bookmarking function is that it is machine-specific, meaning a user can only access their bookmarks from the device on which they are stored. Also, readily sharing the bookmark list with others, or allowing them to discover your bookmark list, was not a standard feature of browser software. Just as an ability to share and find multimedia files like video, audio and images has emerged as part of the Web 2.0/social media "radical present," social bookmarking has become a popular activity online, fuelled by the tagging functionality of the services:

> the act of tagging a resource is similar to categorizing personal bookmarks ... however, these tools have recently increased in popularity as elements of social interaction have been introduced, connecting individual bookmarking activities to a rich network of shared tags, resources, and users [42].

Examples of such services include Delicious, Stumbleupon and Digg. A second function of bookmarking that overlaps with a multimedia is that of rating websites. This ability to rate material provides an incentive to the compilation of bookmarked sites.

7.3 Tensions within a Networked Society

Figure 7.1 demonstrates that the balance of social networks is being disrupted and changed by widespread transformation of media forms. Since the effects of networked social media are felt in almost every facet of life, the digital then becomes inherently political [43]. As Hassan and Thomas explain,

> Questions of governance, of citizenship, of access and equity, of democracy and the nature of political representation in the age of information have all become salient and hotly disputed. New ... technologies are creating new environments, new political fields of contestation where old political questions are being fought anew [43].

There are many examples of opposing models applicable to life in the information age. According to Barbrook, throughout the latter half of the 20th century, the information society has been identified as "a state plan, a military machine, a mixed economy, a university campus, a hippy commune, a free market, a medieval community or a dotcom firm" [44]. Barbrook elaborates further:

> During these five decades, these rival definitions came in and out of fashion as the fortunes of their promoters waxed and waned. Only one principle remained constant throughout. If about nothing else, the rival ideologues agreed that building the Net was making the future society [44].

Advanced capitalist societies have long operated under a "control paradigm" in terms of the study of culture. The growth of general use of the Internet and new forms of media, however, have fuelled a global ideological realignment [45].

7.3.1 *Freedom of Expression*

In the current global political climate, and with the rapid uptake of new forms of technology as outlined in previous sections of this review, information culture is performing a different function, and has undergone a radical paradigm shift. According to McNair, in the new networked information environment, the binary oppositions that have structured debate and writing since the 1970s have become blurred, and instead of debates about left and right, reactionary and progressive, the new divisions are between modernity and medievalism, secularism and religious totalitarianism.

Post-September 11, to an even greater extent than was true before, advocates of left and right find themselves necessarily united against the murderous assertion of misogynistic, homophobic, racist ideas, and in defence of the right to attack and disagree with each other [46].

This stated right (to attack and disagree with each other), McNair says, is closely aligned with the political issue of free speech. This environment is characterized by instability, "in which no elite group, of whatever ideological position, and however firmly anchored in the corridors of power, is insulated" [46]. The widespread political unrest in the Islamic states of the Middle East in 2011 seems to bear out this prediction.

This situation, according to McNair, has been brought about by three factors: new technologies have increased the speed of information flow through digital networks in such a way that it is difficult to exercise any official control over it; the fact that people no longer accord as much deference to people in elite positions and expect them to be more accountable; and that competition puts pressure on everyone to constantly feature up-to-the minute information.

It is important, however, not to afford too much credence to the notion of the Internet as a utopian vision of the future, according to Barbrook, as in his view the convergence of information provision, telecommunications and networked computing can never liberate humanity. In Barbrook's view, the Internet is little more than a useful tool, and should not be "fetishised" over human input in the grand narrative of history. He asserts that at this point in history, "ordinary people" have taken control of information technologies, and this should provide hope that cooperation and participatory democracy can extend from this virtual world into other areas of human life. Barbrook explains this theory further:

> This time, the new stage of growth must be a new civilisation. Rather than disciplining the present, these new futurist visions can be open-ended and flexible. We are the inventors of our own technologies. We can master our own machines. We are the makers of the shape of things to come. We can intervene in history to realise our own interests. Our utopias provide the direction for the path of human progress. Let's be hopeful and courageous when we imagine the better futures of libertarian social democracy [44].

Concern is expressed in the literature about how online information networks are regulated and may be regulated in future [47, 48]. There are of course examples of this happening already—there are Internet "black holes" in Belarus, Burma, China, Cuba, Egypt, Iran, North Korea, Saudi Arabia, Syria, Tunisia, Turkmenistan, Uzbekistan and Vietnam [49], and there are many other countries where Internet access is very poor indeed, including many on the African continent [50].

A policy introduced by the Australian federal government in 2008, to provide a local context, also proposes mandatory Internet filtering for all Australians and has met with substantial opposition. Though legislation is unlikely to be enacted in support of this policy, the current government is still planning to introduce legislation of some description that supports mandatory Internet filtering in Australia [51].

Some commentators see this as one of the key political issues around the growth of information and social and cultural networks. "The move to electronic communication may be a turning point that history will remember . . . so what we think and do today may frame the information system for a substantial period in the future" [46].

This emphasis on information, and the free unhindered flow of it, highlights another important concept in the literature: that the networked digital landscape enables the democratization of information and even culture. According to Barber [29], the free flow of information and communication is essential to the democratic form of government: "Linked together horizontally by a point-to-point medium like the internet, citizens can subvert political hierarchy and nurture an unmediated civic communication" [29].

7.3.2 *Memes*

Digital dissent, or the disruptive use of information networks, is the use of civic participation online to interrupt traditional narratives of mainstream media, using different online forms including blogs, satire and multimedia memes [21]. "Memes" are defined as units of cultural transmission or units of imitation [52]. Dawkins elaborates further on the concept of memes:

> Examples of memes are tunes, ideas, catch-phrases, clothes fashions, ways of making pots or building arches. Just as genes propagate themselves from the gene pool by leaping from body to body via sperms or eggs, so memes propagate themselves in the meme pool by leaping from brain to brain via a process, which, in the broad sense, can be called imitation [52].

A meme, as associated with the networked social media, is a faddish popular phenomenon. This can be used in a dissenting manner as information can spread very quickly on the Internet—another term for it would be viral messaging. The Internet is an ideal forum for such content, dissenting or otherwise, to spread extremely quickly.

7.3.3 *Social Freedom*

There are two suggested dimensions of the relationship between technology and active and responsible citizenship: that new technologies are a key resource in engendering active and informed citizens; and that citizens are increasingly expected to make their own judgments about scientific and technological matters [45]. Ironically, as Barber explains,

> Those who might most benefit from the net's democratic and informational potential are least likely to have access to it, the tools to gain access, or the educational background to take advantage of the tools. Those with access, on the other hand, tend to be those already empowered in the system by education, income and literacy [29].

Barber further says that despite the potential for inequality and abuse, digital networks can offer much assistance in supporting a democratic society:

> A free society is free only to the degree that its citizens are informed and that communication among them is open and informed ... a guarded optimism is possible about technology and democracy, but only if citizen groups and governments take action in adapting the new technology to their needs [29].

There are many examples of "open and informed" communication through the platform of digital networks. Singer and activist Peter Gabriel has set up a website, Witness [53], which documents human rights abuses, and which the literature records as "one benchmark of how to use real-time media and the video camera in an effective way" [47]. Questions and uncertainties about how to progress remain, however: there are fears that "empathy framing, humanitarian values, and searing visual imagery won't be enough to challenge policy-makers" [47].

7.3.4 *Privacy*

This speculation in the literature over privacy includes the notion that government agencies use it to gather information about citizens for some unknown, politically motivated purpose. In fact, the CIA has been using Facebook to recruit potential employees since December 2006 [54]. The FBI are also highly active in their use of social media [55], while Australia's spy organization, ASIO, is using social media applications to recruit would-be intelligence officers [56]. The U.S. Library of Congress has also recently announced that it will archive billions of "tweets" on Twitter—every message sent on Twitter since it launched in 2006, in fact—in a new archive accessible to approved researchers [57].

The accuracy of the information presented on social networking sites must be carefully assessed. On any of these networking sites, there are many profiles with the same name, and also many profiles created by people pretending to be other people. Fletcher details this further:

> Now that MySpace is owned by News Corp, there are lots of spoof profiles claiming to be the News Corp boss Rupert Murdoch. You can also find any number of George Bushes, Tony Blairs and [leader of the British Conservative Party at the time] David Camerons [58].

Despite the easy accessibility of profiles on social networking sites, many people seem to have the view that they are not public, and are shocked at the ease with which personal information can be extracted from them. This is perhaps a symptom of people not being aware that they can adjust their privacy settings to a highly granular level on all of these websites. It is as if the pace of technological change is outstripping the ability of individuals to keep up. Inevitably then any description of social media will be a snapshot of a moment in time and like the science fiction of Gibson of the forecasting of Kurtzweil say more about the present digital environment that it can about the future. Here is a version of that snapshot.

7.3.5 *The Future of Social and Cultural Networks*

The geography of digital social and cultural networks will continue to change and the growth in new technologies will shape this "cybernetic" space. These will in turn have concomitant social, political and cultural ramifications.

The literature states that in 2006, Australians were ranked in the top five nations globally in terms of their Internet use, and that 70.7% of Australians had Internet access in September 2006, compared with 69.3% of people in the United States and 62.5% in the United Kingdom (Internet World Statistics, cited in [59]).

The online medium is beginning to take precedence over the printed word in terms of information supply, and is gaining ground in Australia [59]. In addition, the way people access information through digital networks is constantly evolving, and personalization is seen as a key feature of this evolution.

Personalization tools are part of a move, the literature says, towards a new "configurative" form of literacy, rather than a linear one. In our increasingly interactive networked digital information landscape, which sees us moving from passive consumption to active participation, from interpretive to configurative practices, we have a new relationship to online information [60].

Personalization is both a way to navigate the huge amount of information proliferating on digital information networks, but also allows people to focus on particular subject areas of specific interest. Lasica (2002) describes inclusive personalization as follows:

> True personalisation requires an extra step: a recurring set of interactions between news provider and news consumer that permits you to tailor the news to your specific interests. Imagine a publication made up entirely of articles of special interest to you: stories about your hometown, your college, field of study, hobbies and interests, favourite bands, TV shows and sports teams, along with coupons and discounts for all the stuff you need to buy. Call it the "Daily Me" [61].

Academia is also funding further research into digital information studies. In early 2007, the MIT Media Lab and Comparative Media Studies Program received a four-year, $US5 million grant from the Knight Foundation in the United States to establish the Centre for Future Civic Media. According to its mission statement, the centre will

> develop and study new technical and social systems that allow geographically local communities to share, prioritise, organise, and act on information. Integrating engineering, analysis, and debate, the Centre will take techniques and technologies that have proven so powerful for distributed and virtual communities, and re-envision and re-engineer them to enhance civic engagement at a local level [62].

According to an MIT press release announcing the awarding of the grant, the term "civic media" refers to any form of communication that strengthens the social bonds within a community or creates a strong sense of civic engagement among its residents [63]. Civic media, it says, is more than newsgathering and reporting, and students at MIT are experimenting with everything from technologies designed to be used during protests and civil disobedience, to phone-texting systems that allow them to vote instantly on everyday activities. The Centre, MIT states, is focused on extending and enhancing the reach of these technologies to empower communities, while at the same time generating curricula and "open-source frameworks for civic action":

> Transforming civic knowledge into civic action is an essential part of democracy. As with investigative journalism, the most delicate and important information can often focus on leaders and institutions that abuse the trust of the communities they serve. By helping to provide people with the necessary skills to process, evaluate and act upon the knowledge in circulation, civic media ensures the diversity of inputs and mutual respect necessary for democratic deliberation [63].

7.3.6 *Immersion in Digital Cultural Networks*

Howard Rheingold's book *Smart Mobs: The Next Social Revolution* identifies the ability of online information networks to inspire "moblike" behaviour and display facets of what he calls the "hive mind." Rheingold's explanation of the "scramble system," based on the model of the four-way pedestrian crossing, describes people "performing a complex, collective, ad hoc choreography that accomplishes the opposite of flocking; people coordinate with immediate neighbours to go in different directions" [64]. This scramble system has similarities to the following notion, posited in Pierre Levy's book *Collective Intelligence*:

> Cyberspace provides us with the opportunity to experiment with collective methods of organisation and regulation that dignify multiplicity and variety.... Far from merging individual intelligence into some indistinguishable magma, collective intelligence is a process of growth, differentiation, and the mutual revival of singularities [65].

7.3.7 *Gamers*

Digital games are also an important future consideration in the field of social and cultural networks, not least because some regard them as the emergent cultural form of our time [66]. As participants in digital social and cultural networks, we are all gamers. McKenzie Wark believes this is the case, and bases his book *Gamer Theory* on this premise:

> Welcome to gamespace. It's everywhere, this atopian arena, this speculation sport. No pain no gain. No guts no glory. Give it your best shot. There's no second place. Winner takes all. Here's a heads-up: In gamespace, even if you know the deal, are a player, have got game, you will notice that the game has got you. Welcome to the thunderdome. Welcome to the terrordome. Welcome to the greatest game of all. Welcome to the playoffs, the big league, the masters, the only game in town. You are a gamer whether you like it or not, now that we all live in a gamespace that is everywhere and nowhere. As Microsoft says: Where do you want to go today? You can go anywhere you want in gamespace, but you can never leave it [66].

7.3.8 *Crowdsourcing*

Another increasing trend within social and cultural networks is the concept of media and other organizations utilizing a method known as "crowdsourcing." According to Niles, crowdsourcing is "the use of a large group of readers to report a news story. It differs from traditional reporting in that the information collected is gathered not manually, by a reporter or team of reporters, but through some automated agent, such as a website" [67].

This method is not new; it has its roots in such practices as news organizations asking readers to phone or email in news tips, but requires a great deal of painstaking work to sort and prioritize them. According to Niles [67], journalists need to recognize that journalism is ultimately a social science, and that if they want to utilize crowdsourcing techniques they need to understand the mathematics of social science. Twentieth-century reporting, he says, will seem inadequate when compared to the techniques that journalists will be using in the 21st century to harness the interactivity and scope of the Internet and gather large quantities of data, of the sort that were previously only used in formal social science studies.

We can in fact see this happening now in 2010: online newspapers such as The Guardian are harnessing this Web activity and creating news stories from data sets as Niles describes. According to the editor of the *Guardian*, Alan Rusbridger, reporters around the world are increasingly making it their mission to publish everything, to make data truly free: "The web has given us easy access to billions of statistics on every matter. And with it are tools to visualise that information, mashing it up with different datasets to tell stories that could never have been told before" [68].

7.4 Conclusion

We could easily twist ourselves in knots trying to keep pace with all of the tools and techniques that will continue to emerge from the intersection of information networks and technology. For what it's worth, one approach is to try and focus on the verbs rather than the nouns. Trying to invest heavily in one particular online "brand" simply because it is popular with a target audience at any given moment is a risky strategy—just ask Rupert Murdoch about News Limited's huge dollar investment in MySpace just before YouTube, FaceBook and Twitter exploded in popularity. When working to develop social media awareness and understanding of networked communication, the brands (the nouns) are important innovators and market leaders, but it is the activities they allow (the verbs) that can guide a more strategic approach to their possible application:

It's easy to get seduced by the "nouns" ...	**But look to the verbs for networks ...**
Facebook	Searching for and verifying information
Youtube	Networking (meeting new people)
Twitter	Social graphing (tracing existing friendships)
MySpace	Creating, re-creating and sharing media
Google	Collaborating on content, solving problems
Yahoo!	Discussing common interests, connecting
Wikipedia	Playing games for fun, or
Hotmail	Exploring through playful behaviour
Blogger	Organising activities
Flickr	Buying or supplying goods and services
Digg	Learning and teaching
Bebo	
LinkedIn	

As we are living through an unprecedented period of fast-moving change, "from a world where computers and computer-based media were peripheral, to one where they are central" [69]. Our lives and our sense of the world—culturally and socially—are being transformed through the use of networked communication. A focus on the verbs or network functions seems a more useful strategy than a focus on the ephemeral nouns or applications that are currently in favour.

A theme that runs through much of the literature of networked communication is the notion that a profound social and cultural revolution or paradigm shift has occurred, and continues to develop due to the emerging digital forms of networked communication, distribution and production. The advent of social media as we know it today has been brought about by the convergence of computing and media technologies and devices. This development is in the process of creating a networked society connected through the ubiquitous new digital forms of cascading social behaviour and producing the unthinkable complexity that Gibson saw in the "unthinkable present" back in 1984.

References

1. Gibson, W. (1999). Talk of the nation. Available at http://www.npr.org/templates/story/story.php?storyId=1067220. Accessed January 2010.
2. Gibson, W. (1984). *Neuromancer* (HarperCollins, London).
3. Dewdney, A., and Ride, P. (2006). *The New Media Handbook* (Routledge, Abingdon, Oxon).
4. Jackson, D. S. (1995). Extra! Readers talk back. *Time.*
5. Doctorow, C. (2009). Cory Doctorow: radical presentism. Tin House Blog. Available at http://tinhousebooks.com/blog/?p=410. Accessed January 2010.
6. Kurzweil, R. (1990). *The Age of Intelligent Machines* (The MIT Press, Cambridge, MA).
7. Prensky, M., and Sapiens H. (2009). Digital: from digital immigrants and digital natives to digital wisdom. *Innovate* 5(3).
8. Wayt Gibbs, W. (1997). Gordon E. Moore, Part 2. *The Scientific American.*
9. Deuze, M. (2006). Participation, remediation, bricolage: considering principal components of a digital culture. *Information Society* 22(2), pp. 63–75.
10. Jenkins, H. (2006). *Convergence Culture: Where Old and New Media Collide* (New York University Press, New York).
11. Naughton, J. (2006). Blogging and the emerging media ecosystem. Reuters Fellowship, Oxford University.

12. Bolter, J. D., and Grusin, R. (2000). *Remediation: Understanding New Media*, 1st ed. (The MIT Press, Boston).
13. Harris, R. (2009). Social media ecosystem mapped as a wiring diagram. Twitter Thoughts Availableat http://www.twitterthoughts.com/social-media-news-analyses/2009/9/3/social-media-ecosystem-mapped-as-a-wiring-diagram.html. Accessed 2011.
14. Li, C. (2007). Social technographics: mapping participation in activities forms the foundation of a social strategy. Forrester Research.
15. Pavlik, J. V. (2004). A sea-change in journalism: convergence, journalists, their audiences and sources. *Convergence* 10(4), pp. 21–29.
16. Wall, M. (2005). "Blogs of war": weblogs as news. *Journalism* 6(2), pp. 153–172.
17. Matheson, D. (2004).Weblogs and the epistemology of online news: some trends in online journalism. *New Media & Society* 6(4), pp. 443–468.
18. Turkle, S. (1995). *Life on the Screen: Identity in the Age of the Internet* (Simon and Schuster, New York).
19. Ibrahim, Y. (2006). Weblogs as personal narratives: displacing history and temporality. *M/C: A Journal of Media and Culture*. 9.
20. Allan, S. (2006). *Online News: Journalism and the Internet* (Open University Press, Maidenhead, Berkshire).
21. Boler, M. (2006). The transmission of political critique after 9/11: "A new form of desperation"? *M/C: A Journal of Media and Culture* 9.
22. Gibson, O. (2003). Spin caught in a web trap. *The Guardian*.
23. Huberman, B. A., Romero, D. M., and Wu, F. (2009). Social networks that matter: Twitter under the microscope. *First Monday* 14(1).
24. Java, A., et al. (2007). Why we twitter: understanding microblogging usage and communities. International Conference on Knowledge Discovery and Data Mining, San Jose, California. ACM.
25. Kwak, H., et al. (2010). What is Twitter, a social network or a news media? 19th International World Wide Web Conference, Raleigh, NC.
26. Viégas, F. B. (2005). Bloggers' expectations of privacy and accountability: an initial survey. *Journal of Computer-Mediated Communication* 10(3).
27. Meikle, G. (2003). Indymedia and the new net news. *M/C: A Journal of Media and Culture* 6(2).
28. Pool, I. d. S. (1983). *Technologies of Freedom: On Free Speech in an Electronic Age*. (Harvard University Press, Cambridge, MA).

29. Barber, B. R. (1988). Pangloss, Pandora or Jefferson? Three scenarios for the future of technology and strong democracy, in *The New Media Theory Reader*, R. Hassan and J. Thomas, eds. (Open University Press, Maidenhead, Berkshire), pp. 163–203.
30. Bentham, J. (1995). *The Panopticon Writings*, M. Bozovic, ed. (Verso, London), pp. 29–95.
31. Brignall III, T. (2002). The new Panopticon: the Internet viewed as a structure of social control. *Theory & Science* 3.
32. Tseng, E., and Eischen, K. (2003). The geography of cyberspace. *M/C: A Journal of Media and Culture* 6(4).
33. Barnes, S. B. (2006). A privacy paradox: social networking in the United States. *First Monday* 11(9).
34. Acquisti, A., and Gross, R. (2006). Imagined communities: awareness, information sharing, and privacy on the Facebook. 6th Workshop on Privacy Enhancing Technologies (Springer Verlag).
35. Jones, T. (2010). Facebook's "Evil Interfaces."
36. Opsahl, K. (2010). Facebook's Eroding Privacy Policy: A Timeline.
37. Bogost, I. A professor's impressions of Facebook. 2007.
38. Leuf, B. and Cunningham, W. (2001). *The Wiki Way: Quick Collaboration on the Web* (Addison-Wesley, Boston).
39. Wikipedia. Wabi sabi. Available at http://en.wikipedia.org/wiki/Wabi-sabi. Accessed August 15, 2007.
40. Reagle, J. M. (2010). *Good Faith Collaboration: The Culture of Wikipedia* (MIT Press, Cambridge, MA).
41. Tapscott, D., and Williams, A. D. (2006). *Wikinomics: How Mass Collaboration Changes Everything* (Portfolio, New York).
42. Marlow, C., et al. (2006). HT06, Tagging Paper, Taxonomy, Flickr, Academic Article, To Read. Hypertext '06 Conference, ACM, Odense, Denmark.
43. Hassan, R., and J. Thomas, eds. (2006). *The New Media Theory Reader* (Open University Press, Maidenhead, Berkshire).
44. Barbrook, R. (2007). *Imaginary Futures: From Thinking Machines to the Global Village* (Pluto Press, London).
45. Barry, A. (2001). On interactivity, in *The New Media Theory Reader*, R. Hassan and J. Thomas, eds. (Open University Press, Maidenhead, Berkshire), pp. 163–187.
46. McNair, B. (2003). From control to chaos: towards a new sociology of journalism. *Media Culture Society* 25(4), pp. 547–555.

47. Burns, A. (2003). The worldflash of a coming future. *M/C: A Journal of Media and Culture* 6(2).
48. Pool, I. d. S. (2003). A shadow darkens, in *The New Media Theory Reader*, R. Hassan and J. Thomas, eds. (Open University Press, Maidenhead, Berkshire), pp. 19–26.
49. Reporters without borders. Available at http://www.rsf.org/rubrique.php3?id_rubrique=273. Accessed August 24, 2007.
50. Alden, C. (2004). For most Africans, Internet access is little more than a pipe dream. *Online Journalism Review*.
51. Conroy, S. S. (2009). Measures to improve safety of the internet for families. 2009 Available at http://www.minister.dbcde.gov.au/media/media_releases/2009/115. Accessed April 21, 2010.
52. Dawkins, R. (1976). *The Selfish Gene*: (Oxford University Press, New York).
53. Balnaves, M., Donald, S. H., and Shoesmith, B. (2009). *Media Theories and Approaches: A Global Perspective* (Palgrave Macmillan, London).
54. Bruce, C. (2007). CIA gets in your Face(book). *Wired*.
55. The portable FBI: our newest social media initiatives. Available at http://www.fbi.gov/page2/jan10/social_010710.html. Accessed April 21, 2010.
56. Rodgers, E. (2010). ASIO hunts for new recruits on Facebook. *ABC*.
57. Anderson, N. (2010). Why the Library of Congress cares about archiving our Tweets. *Ars Technica*.
58. Fletcher, K. (2007). Why blogs are an open door. *British Journalism Review* 18(2), pp. 41–46.
59. Cowden, G. (2007). Online news: patterns, participation and personalisation. *Australian Journalism Review* 29(1), pp. 75–85.
60. Moulthrop, S. (2004). From work to play: molecular culture in the time of deadly games, in *First Person: New Media as Story, Performance and Game*, N. Wardrip-Fruin and P. Harrigan, eds. (The MIT Press, Cambridge, MA), pp. 56–69.
61. Lasica, J. (2002). The promise of the Daily Me—From My News to digital butlers: an in-depth look at the different flavors of personalisation. *Online Journalism Review*.
62. Weinberger, D. (2007). *Everything Is Miscellaneous: The Power of the New Digital Disorder* (Henry Holt and Company, New York).
63. MIT announces a new Centre for Future Civic Media. 2007. Available at http://civic.mit.edu/?p=4. Accessed September 14, 2007.

64. Rheingold, H. (2002). *Smart Mobs: The Next Social Revolution* (Basic Books, New York).
65. Levy, P. (1997). *Collective Intelligence: Mankind's Emerging World in Cyberspace* (Basic Books, New York).
66. Wark, M. (2007). *Gamer Theory* (Harvard University Press, Cambridge).
67. Niles, R. (2007). A journalist's guide to crowdsourcing. *Online Journalism Review.*
68. Rusbridger, A. (2010). Free the facts. Data blog: Facts are sacred.
69. Carroll, J., (2007). Trying to reach that elusive younger generation. *The Bulletin of Public and Corporate Communication* 2007(4).

Author Index

Subject Index